CHEMISTRY RESEARCH AND APPLICATIONS

VEGETABLE OIL

PROPERTIES, USES AND BENEFITS

CHEMISTRY RESEARCH AND APPLICATIONS

Additional books in this series can be found on Nova's website
under the Series tab.

Additional e-books in this series can be found on Nova's website
under the eBooks tab.

CHEMISTRY RESEARCH AND APPLICATIONS

VEGETABLE OIL

PROPERTIES, USES AND BENEFITS

BRITTANY HOLT
EDITOR

nova publishers
New York

Library of Congress Cataloging-in-Publication Data

ISBN: 978-1-63485-128-2

Published by Nova Science Publishers, Inc. † New York

CONTENTS

PREFACE

Chapter 1 – Vegetable oils are hydrophobic compounds, such as tryacylglicerols and essential oils extracted from plants that have been used by humans for centuries in many different areas. The use of isolated enzymes and microorganisms applied to vegetable oils has been shown to be very interesting from an industrial viewpoint, due to the broad variety of products that may be obtained from it. As an example, biodiesel can be obtained through transesterification of vegetable oils, such as soybean and rapeseed oils, catalyzed by free and immobilized lipases from different microorganisms, including *Candida antarctica* and *Pseudomonas fluorescens,* with a conversion rate of more than 90% according to some studies. This is of great interest considering the current scenario of biofuels and environmental consciousness. Lipases from microorganisms such as *Mucor miehei* and *Candida antarctica* have also been studied for the production of structured lipids through interesterification reactions, using vegetable oils and any other oil of interest to be incorporated as substrates such as conjugated linoleic acid. These modified lipids are attractive since they present improved therapeutic or technological properties compared to natural vegetable oils. Furthermore, lipases can be used as biocatalysts in the production of useful biodegradable compounds, using vegetable oils as substrates, including the production of 1-butyl oleate from direct esterification of butanol and oleic acid, to decrease the viscosity of biodiesel in winter use. Another example is the mixture of 2-ethyl-1-hexyl esters that can be obtained efficiently by enzymatic transesterification from rapeseed oil fatty acids, for use as a solvent. On the other hand, compounds of essential oils can be biotransformed by microorganisms to obtain high value-added compounds. *Penicillium* sp. has been studied for the biotransformation of *R*-(+)-limonene into *R*-(+)-α-terpineol, a compound with a floral and typically lilac odor that can be used as an aroma compound in many products. Additionally, metabolic engineering of plants to obtain high levels of oils with industrially desired properties, such as reduced viscosity, freezing point and calorific value, has been recently studied. Therefore, vegetable oils and biocatalysis represent a promising strategy in overcoming environmental issues and being economically viable. Many efforts must be made in order to understand the biological conversion and the development of new technologies. This chapter focuses on vegetable oils with particular emphasis on the use of enzymes and microorganisms for their modification in order to produce industrially relevant compounds.

Chapter 2 – In recent years the production of vegetable oils (sunflower, linseed, rapeseed, soybean, olive, etc.) has significantly increased, oils being considered a major economic resource, mainly used in the food industry or as raw materials for the biodiesel production, as well as for the oleochemical and pharmaceutical industries.

Authentication of vegetable oils refers mainly to two aspects. On one hand, it is related to establishing conformity according to label - in terms of plant species, part of the plant (e.g., discrimination between palm oil and palm kernel oil), geographical origin, crop year, processing technique (differentiation of minimum processed oils such as cold pressed oils from refined oils). On the other hand, detection of adulteration of expensive vegetable oils and fats such as olive oil or cocoa butter with cheaper oils is another important issue concerning oils authentication.

In the last decade, researchers in the food quality control field, showed an increased interest to establish methods for products authentication. This trend results from the consumers enhanced demand for quality products. Vegetable oils authentication by means of establishing the geographical origin of oil seeds goes with the new trend of protecting the local food products. Protected designation of origin (PDO), protected geographical indication (PGI) and Traditional Specialty Guaranteed (TSG) are geographical indications defined in European Union Law to protect regional foods and to eliminate the dishonest practices. Those laws are inflicted in the European Community since 1992 and gradually extended to the non-UE countries by various joint agreements.

The fatty acid composition of vegetable oils is the main factor influencing their nutritional value and properties. In recent years, new and efficient methods for the determination of the fatty acids profile of different kind of samples (e.g., vegetable oils, lipid extracts from different vegetable or animal tissues) have evolved, being partially fed by the general trend in the chemical literature concerning the food authenticity assessment. Among the most envisaged physical methods of analysis, vibrational spectroscopy (NIR, FT-IR and Raman), nuclear magnetic resonance (^{1}H-, ^{13}C-NMR) and chromatographic techniques (GC, HPLC) can provide structural and compositional information useful for vegetable oils authentication. Special attention has been focused on ^{1}H-NMR spectroscopy which leads to a global fatty acids profile (in terms of linolenic, linoleic, mono-unsaturated and saturated fatty acids) in a non-invasive manner and mild analysis conditions which do not require chemical transformations prior to/during analysis, as compared to chromatographic techniques. Moreover, spectra (both NMR or vibrational) can be considered as fingerprints of samples taken into analysis, thus being helpful for authentication purposes.

Authentication of a food product is almost impossible without the use of statistical methods applied to data bases obtained from genuine samples. Among these methods, multivariate statistical analysis - *Principal Component Analysis* (*PCA*), *Discriminant Analysis* (*DA*) - have proven suitable for vegetable oils authentication.

In this chapter, the authors aim to investigate the potential of physical methods (NMR or vibrational spectroscopies and chromatography) coupled with statistical data processing to assess vegetable oils authentication issues such as establishing conformity according to label and detection of adulteration.

Chapter 3 – Declining global fossil fuel reserves due to escalating consumption and associated environmental pollution created an urge for investigating into production of renewable and environmentally friendly biofuels. Biofuels such as biodiesel are most sought because they can be produced in liquid form suitable for most transportation needs. Moreover, biodiesel can be produced from a variety of oil-rich feedstock and even from waste cooking oils, animal fat and other microbial lipids. This book chapter provides an overview of the oil feedstock suitable for biodiesel production and their characteristics. The chemistry and process of biodiesel production are discussed in detail.

Chapter 4 – Limited reserves of fossil fuels as well as the growing concern for the environment, has led to a worldwide search for renewable energy sources, among which biodiesel, a mixture of fatty acid methyl esters (FAME), is one of the most perspective alternative fuels since it is a non-toxic and can be produced from different renewable sources through simple cost-effective alcoholysis, while being compatible with existing infrastructures. Vegetable oils, as renewable in nature and environmentally friendly, with a possibility to be produced on a large scale, represent a promising feedstock for biodiesel production. In this chapter, a comprehensive review of different vegetable oils as a feedstock for biodiesel synthesis is reported, including edible and non-edible oils, as well as waste vegetable oils.

Selection of feedstock for biodiesel production mainly depends on the specific conditions and circumstances in some region (climate, presence of certain crops, the economic development of a country, etc.). Various fatty acid compositions of triacylglycerols directly determine the quality and fulfillment of the standards of biodiesel. One of the crucial points which determine technology route for biodiesel synthesis is content of free fatty acids (FFA) which might be present in vegetable oils, as well as the presence of water and other compounds. Also, many analysis performed in the past have shown that the production cost of biodiesel is mainly determined by the price of used feedstock, which represents 70–80% of total production costs of biodiesel. Currently more than 95% of feedstock comes from edible oils since they are easily accessible, consists mainly of triacylglycerols, whereby the properties of biodiesel produced from these oils are suitable to be used as diesel fuel substitute. Most commonly used edible oils for biodiesel production are rapeseed, soybean, sunflower and palm. However, for economic and social reasons, in recent years research and development of biodiesel production has focused on other sources of triacylglycerols, in order to replace edible oils by lower-cost non-edible plant oils and the waste cooking oils, feedstocks that are unsuitable for human consumption.

Properties of different oils and biodiesel obtained from them as well as the technologies suitable for biodiesel synthesis are compared in this chapter. The well known fact is that the conventional and to-day widely applied homogeneous method of biodiesel synthesis is sensitive to the presence of impurities in the oil, primarily the presence of FFA and water. Furthermore, biodiesel synthesis is followed by creation of large amount of wastewater produced during neutralization of catalyst and purification of final product. The drawbacks of a homogeneous process can be avoided by applying technologies based on utilization of heterogeneous catalyst or by application of the non-catalytic supercritical process of biodiesel synthesis. These technologies for biodiesel production were also analyzed and compared.

Chapter 5 – Ozonated vegetable oils have demonstrated promising results for clinical application, and they have been the focus of great pharmaceutical interest to treat dermatological disorders, such as infections of skin ulcers and chronic wounds. There are reports of these products as effective to heal refractory wounds, where conventional treatments and available medications prove ineffective. In fact, in some European countries, such as Germany, they can be obtained on prescription from pharmacies. Countries such as Cuba have developed commercial ozonated oils, and they have been successfully tested to treat many diseases. Cuba is one of the pioneers in the implementation of this therapy in Public Health Services for over 22 years. Ozone reacts with the double bonds of unsaturated fatty acids of vegetable oils, providing stable ozonation products, mainly ozonides, hydroperoxides and polyperoxides (depending on reaction conditions) with therapeutic potential. Several studies have demonstrated their antimicrobial and antifungal activity, as well as their role as wound healing modulators, showing no cytotoxicity when tested against NIH/3T3 murine fibroblast cells. Simple analytical techniques such as peroxide value, iodine value and viscosity determination have been extensively used for characterization of products, together with spectroscopic techniques of NMR (^{1}H and ^{13}C) and infrared, chromatography and thermal analysis (TG/DTG -DSC). This chapter aims to highlight recent contributions to the production, characterization and biological activities of ozonated vegetable oils.

Chapter 6 – Virgin coconut oil (VCO) is a product that can be produced from fresh coconut meat, milk, or residue. Over the years, it has become known as a popular functional food oil. It is considered to be the newest, high-value coconut product, very much sought for its human, nutraceutical benefits, as well as a functional food. Its increasing popularity can be attributed to numerous studies showing its beneficial effects. Several studies have investigated the pharmacological properties of VCO including anti-inflammatory, analgesic, antipyretic, anti-oxidant, anti-stress, and antimicrobial properties. Furthermore, other studies have also investigated the bone loss prevention as well as cardioprotective.

In: Vegetable Oil
Editor: Brittany Holt

ISBN: 978-1-63485-128-2

Chapter 1

AN OVERVIEW ON VEGETABLE OILS AND BIOCATALYSIS

Ricardo Rodrigues de Melo[1,2], Gustavo Pagotto Borin[1,3], Guilherme Keppe Zanini[1,3], Antonio Adalberto Kaupert Neto[1,3], Bruna Soares Fernandes[1] and Fabiano Jares Contesini[1,*]

[1]Brazilian Bioethanol Science and Technology Laboratory (CTBE), Brazilian Centre of Research in Energy and Materials (CNPEM), Campinas, São Paulo, Brazil

[2]Faculty of Food Engineering, University of Campinas, Campinas, Brazil

[3]Institute of Biology, University of Campinas, Campinas, Brazil

ABSTRACT

Vegetable oils are hydrophobic compounds, such as tryacylglicerols and essential oils extracted from plants that have been used by humans for centuries in many different areas. The use of isolated enzymes and microorganisms applied to vegetable oils has been shown to be very interesting from an industrial viewpoint, due to the broad variety of products that may be obtained from it. As an example, biodiesel can be obtained through transesterification of vegetable oils, such as soybean and rapeseed oils, catalyzed by free and immobilized lipases from different microorganisms, including *Candida antarctica* and *Pseudomonas fluorescens,* with a conversion rate of more than 90% according to some studies. This is of great interest considering the current scenario of biofuels and environmental consciousness. Lipases from microorganisms such as *Mucor miehei* and *Candida antarctica* have also been studied for the production of structured lipids through interesterification reactions, using vegetable oils and any other oil of interest to be incorporated as substrates such as conjugated linoleic acid. These modified lipids are attractive since they present improved therapeutic or technological properties compared to natural vegetable oils. Furthermore, lipases can be used as biocatalysts in the production of useful biodegradable compounds, using vegetable oils as substrates, including the production of 1-butyl oleate from direct esterification of butanol and oleic

* Corresponding author: Fabiano Jares Contesini e-mail: fabiano.contesini@gmail.com.

acid, to decrease the viscosity of biodiesel in winter use. Another example is the mixture of 2-ethyl-1-hexyl esters that can be obtained efficiently by enzymatic transesterification from rapeseed oil fatty acids, for use as a solvent. On the other hand, compounds of essential oils can be biotransformed by microorganisms to obtain high value-added compounds. *Penicillium* sp. has been studied for the biotransformation of *R*-(+)-limonene into *R*-(+)-α-terpineol, a compound with a floral and typically lilac odor that can be used as an aroma compound in many products. Additionally, metabolic engineering of plants to obtain high levels of oils with industrially desired properties, such as reduced viscosity, freezing point and calorific value, has been recently studied. Therefore, vegetable oils and biocatalysis represent a promising strategy in overcoming environmental issues and being economically viable. Many efforts must be made in order to understand the biological conversion and the development of new technologies. This chapter focuses on vegetable oils with particular emphasis on the use of enzymes and microorganisms for their modification in order to produce industrially relevant compounds.

1. INTRODUCTION

Vegetable oils are mainly fluid hydrophobic compounds (at room temperature) obtained from milled grains from different plants such as sunflower, canola, soybean, Jatropha, palm, rapeseed, peanut, cottonseed (Ondul et al. 2015). These oils are composed by lipids, mainly triacylglycerols with different fatty acids ramifications. The most common are caprylic, capric, lauric, myristic, palmitic, palmitoleic, stearic, oleic, linoleic, α-eleostearic, ricinoleic and vernolic. A higher melting point is related to the amount of carbons present in the fatty acid chain, and also a lower number of double bonds between carbons and trans configuration of these bonds (Xia and Larock 2010). Other hydrophobic compounds are also present in considerable amounts in several vegetables such as sterol, carotene, phospholipids and tocopherol.

Palm oil leads the worldwide market of edible vegetable oil. It consists over 95% of mixtures of tryacylglicerols, and minor amount of other classes of lipids, such as sterols, tocopherols, carotenoids, and phospholipids (Sambanthamurthi et al. 2000), very interesting compounds for several industries. According to Cheong et al. (Cheong et al. 2014) these minor constituents show a broad qualitative and quantitative composition depending of the vegetable species and their oil extraction. These minor constituents can have pro oxidative (free fatty acids) or antioxidant (tocopherols, carotenoid and phospholipids) effects that can be maximized by biotransformation increasing shelf life and miscibility in water or in other solvent, changing properties and other effects.

One alternative to process the aforementioned vegetable oils is by biocatalysis or biotransformation using enzymes or microbial cells to produce useful and enriched bioproducts that can convert the plant lipids to aimed compounds. Biotransformation or biocatalysis is a modification of organic compounds via chemical reactions catalyzed by cellular enzymes. Common chemical reactions include oxidation, reduction, hydrolysis, hydroxylation, isomerization, and glycosylation. Additionally, similar process using chemical synthesis often results in environmentally unfriendly processes with lack of substrate selectivity, causing undesirable reactions rising the downstream process costs (Molina et al. 2013).

One of the first patent and technological advance in biotransformation of vegetable oil was developed by Fuji Oil, in 1979, using lipase in the transesterification or acidolysis of cheap oils, tristearin or stearic acid, as acyl donor (Biermann et al. 2011). Other examples are the synthesis of biodiesel using triacylglycerols as substrates and microbial lipases as catalysts (Tan et al. 2010a). Another example is the production of α-terpineol using *Penicillium digitatum* DSM 62840 cells and D-limonene as substrate (Abraham et al. 1986).

Modified oils present several relevant industrial applications to be explored such as their utilization as lubricants, surfactants, emulsifiers and biopolymers (Figure 1). They also represent an excellent source for the obtainment of valuable compounds such as unsaturated fatty acids, phytosterols, squalene, pigments, antioxidants, vitamins, waxes, glycolipids and lipoproteins. They can also serve as appropriate sources for applications in other fields such as food, pharmaceutical and environmental. Some value-added fatty acids (polyunsaturated fatty acids, conjugated fatty acids, hydroxy, keto, epoxy, branched and cyclic fatty acids), biosurfactants based on functional link between carbohydrates and selected lipid fractions (glycolipids, glycosylated and acylglycosylated sterols) and flavor compounds, have been receiving a lot of attention (Certik 2008).

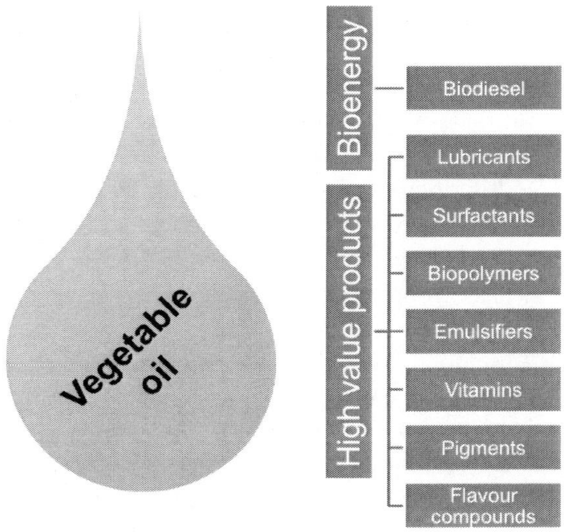

Figure 1. Different applications of products obtained from vegetable oils modification using biocatalysis and biotransformation.

In spite of these high added value bioproducts obtained from plant lipids and their modifications, vegetable oils are vastly applied in the production of renewable energy. As previously cited in the examples of biocatalysis routes, the most common application is the production of biodiesel which requires a highly efficient fermentation process to make it more profitable. The bottleneck in the production of biodiesel is related to the limitation of the enzymes used nowadays, that raises the cost of the production of this kind of biodiesel. Many different approaches have already been used focusing on lipases and its several application modes. Biodiesel production targeting the lowering of costs in enzyme production has improved a lot recently (Teixeira et al. 2014).

A different alternative to obtain modified plant lipids that can be industrially relevant is the genetic engineering of the plants in order to genetically modify routes of lipid synthesis. With that, different compounds can be obtained, such as lipids with different properties from the lipids produced before genetic modification of the plant (Certik 2008, Dyer and Mullen 2008). This review focuses on the modification of vegetable oils using enzymes or microbial cells in order to obtain high value-added products. The use of genetic engineering tools applied to plants to obtain specific modified lipids is also discussed.

2. Biodiesel Production from Vegetable Oils

Since the industrial revolution, different forms of energy have become essential for human beings and to maintain economic growth. In the past few decades, mainly petroleum-based liquid fuels have played an important role in fulfilling this energetic demand (Ashraful et al. 2014). According to BP Statistical Review of World Energy, globally, we consumed in 2014 the equivalent of 13541 million tons of oil (Mtoe). The estimate is 122% higher than in 1973. If this rate continues to increase, oil deposits will be exhausted in 2052 (Pal et al. 2015). Another problem related to the exhaustive use of fossil fuels is the release of significant quantities of greenhouse gas (GHG) emissions resulting in global warming (Kumar and Sharma 2015).

Consequently, the world is moving towards an energetic crisis. This situation has enticed many countries to research and develop alternative sources to substitute petroleum fuels (Demirbas 2009; No 2011; Silitonga et al. 2013). Among the alternative sources, biodiesel emerges as an important and promising resource. The Renewable Fuel Standard program published by the United States Environmental Protection Agency describes that the goal for 2022 will be 36 billion gallons of biofuels, with an estimated production of at least 1 billion gallons of biodiesel (Christopher et al. 2014).

Considerable attention on biodiesel is due its important characteristics; it emits less pollutants, is environment friendly and non-toxic, and biodegradable (Agarwal 2007; Ahmad et al. 2011; Lapuerta et al. 2008). Biodiesel is a mixture of monoalkyl esters of long-chain fatty acids, preferentially methyl and ethyl esters, derived from renewable sources such as edible or non-edible vegetable oils, waste obtained from vegetable oils, and animal fats (Knothe 2010).

Vegetables or plant oils are among the most important renewable resources used in biodiesel production. They have low impact on the environment and do not contribute to the global carbon dioxide (CO_2) liberation, and economic benefits are the main reasons for the use of vegetable oils for biodiesel synthesis (Mosarof et al. 2015). Therefore, vegetable oils as fuels have been analyzed extensively in recent years (Janaun and Ellis 2010; Mofijur et al. 2015; Mohsin et al. 2014). Mosarof et al. (2015) demonstrated the biodiesel production from palm oil as one important fuel alternative (Mosarof et al. 2015). Other studies on alternative fuels have investigated the use of soybeans (Haas 2005), jatropha (Kuo et al. 2015), canola (Zou and Atkinson 2003), sunflower (Calero et al. 2014), cotton seed (Köse et al. 2002), and corn oils as a promise for biodiesel production (Zhang et al. 2011).

The most common approaches for biodiesel synthesis in the world is carried out by chemical processes, using acids and alkalis catalysts, such as H_2SO_4 and NaOH, because of its high yield production and economic viability for the industrial scale (Abdelmoez and Mustafa 2014; Sharma et al. 2008). However, chemical processes present some disadvantages, such as high energy consumption, formation of undesirable by-products difficult to separate, large volume of waste, and the necessity of removal of the chemical catalyst, since it can damage engine parts (Abbaszaadeh et al. 2012; Meher et al. 2006; Vyas et al. 2010).

Contrary to this process, enzyme-catalyzed reactions have proven to be an interesting alternative tool for replacing conventional chemical processes for industrial-scale biodiesel production. Lipases, glycerol ester hydrolases (E.C.3.1.1.3), are enzymes that catalyze hydrolysis of esters, particularly long-chain tryacylglicerols, producing free fatty acids, di- and monoglycerides, and glycerol. However, in organic conditions the lipases can catalyze esterification, interesterification, acidolysis, aminolysis, and alcoholysis reactions with high specificity and selectivity (Ribeiro et al. 2011; Sharma and Kanwar 2014).

Enzymatic process used lipases are able to catalyze both esterification and transesterification reactions, which allows the production of biodiesel from cheaper raw materials with high free fatty acids and water content (Ribeiro et al. 2011). Furthermore, lipases applied in the biodiesel production process generate less waste and effluents, making it more environment friendly. The use of lipases is also responsible for generating lower energy expenditure, since enzymatic catalysis is carried out at mild conditions of temperature and pressure (Robles-Medina et al. 2009).

The most used lipases in biodiesel synthesis are the lipases obtained from *Candida antarctica* (Royon et al. 2007), and *Candida rugosa* (Kuo et al. 2015). Lipases obtained from filamentous fungi, such as *Rhizomucor miehei* lipase (Oliveira and Rosa 2006), *Rhizopus oryzae* lipase (Pizarro and Park 2003), *Thermomyces lanuginosus* lipase (Oliveira et al. 2005), and *Penicillium expansum* lipase (Zhang et al. 2011) have also been studied. Table 1 shows several lipases applied in biodiesel production, with their yields and oils used.

Table 1. Application of free lipases in biodiesel production

Oil	Lipase origin	Acyl acceptors	Maximum yield (%)	Reference
Jatropha oil	*Candida rugosa*	Methanol	95.3	(Kuo et al. 2015)
	Idiomarina sp. W33	Methanol	84.0	(Li and Yu 2014)
Cottonseed oil	*Candida antarctica*	Methanol	97.0	(Royon et al. 2007)
Sunflower oil	*Candida antarctica*	Ethyl acetate	63.3	(Kim et al. 2007)
	Rhizomucor miehei	Methanol	95.5	(Oliveira and Rosa 2006)
	Thermomyces lanuginosus	Ethanol	16-35	(Oliveira et al. 2005)
Corn oil	*Penicillium expansum*	Methanol	86.0	(Zhang et al. 2011)
Palm oil	*Rhizopus oryzae*	Methanol	55.0	(Pizarro and Park 2003)
Soybean oil	*Candida antarctica*	Methanol	93.4	(Talukder et al. 2009)
	Pseudomonas fluorescens	Methanol	90.0	(Kaieda et al. 2001)
	Pseudomonas cepacia	Methanol	>80.0	(Kaieda et al. 2001)
	Candida rugosa	Methanol	90.0	(Kaieda et al. 2001)

Table 2. Application of immobilized lipases in biodiesel production

Oil	Lipase origin	Support used	Yield (%)	Reference
Canola oil	*Alcaligenes* sp.	CLEA-P QL/Crosslinking	>75	(Soler et al. 2016)
	Candida antarctica (CALB)	Epoxy-functionalized silica/Covalent bonding	68.0	(Babaki et al. 2015)
	Thermomyces lanuginosus	Epoxy-functionalized silica/Covalent bonding	98.0	(Babaki et al. 2015)
	Rhizomucor miehei	Epoxy-functionalized silica/Covalent bonding	45.0	(Babaki et al. 2015)
	Candida antarctica Lipase A	Chitosan beads	60.0	(Aybastıer and Demir 2014)
Jatropha oil	*Candida cylindracea*	Functionalized activated carbon support	78.0	(Kabbashi et al. 2015)
	Idiomarina sp. W33	Amberlite IRA-93/Adsorption	91.0	(Li and Yu 2014)
	Rhizopus oryzae	Polyvinyl alcohol-alginate	87.1	(Zarei et al. 2014)
	Chromobacterium viscosum	Celit 545/Adsorption	92.0	(Shah et al. 2004)
Sunflower oil	*Rhizomucor miehei*	Macroporous anion exchange resins	88.6	(Calero et al. 2014)
	Candida rugosa	Celit 545/Adsorption	>80	(Sagiroglu 2008)
	Candida sp. 99–125	Textile membrane/Absorption	80.6	(Lu et al. 2009)
	Candida antarctica Novozyme 435	Acrylic resin beads	92.7	(Modi et al. 2007)
Soybean oil	*Candida rugosa*	Magnetic chitosan microspheres/Covalent bonding	87.0	(Xie and Wang 2012)
	Candida antarctica	Styrene-divinylbenzene beads	99.0	(Poppe et al. 2013)
Palm oil	*Thermomyces lanuginosus*	Macroporous ion-exchange resin	79.0	(Basri et al. 2013)
	Candida antarctica Novozyme 435	Macroporous acrylic resin	92.0	(Talukder et al. 2009)

Although enzymatic reactions using lipases have significant yield in conversion of vegetable oils into biodiesel, current market costs of lipases are still the main limitations that avoid their use in large-scale processes. Estimates suggest that the cost of commercial lipase (Novozym 435) and NaOH is US$ 0.14 and US$ 0.006 per kilogram of ester, respectively (Daiha et al. 2015). A possible solution to reduce the process costs incurred is the use of immobilized lipases. The use of immobilized lipases may improve the development of commercial scale processes, favoring biotechnological processes based on their numerous advantages over the chemical processes (Poppe et al. 2015; Tan et al. 2010b). Therefore, lipases immobilization can promote repeated use of the biocatalysts, reducing operation costs.

Moreover, immobilization becomes a very important step, since it allows more tolerance by the enzyme, and the separation of products is easier (Bajaj et al. 2010).

Many supports have been applied for the lipases immobilization. Generally, these techniques involve traditional methods such as; adsorption, entrapment, covalent binding, and cross-link (Tan et al. 2010b). Industrially, immobilization of enzymes can stimulate the use of the lipases. Immobilized *Candida antarctica* (CALB), *Thermomyces lanuginosus*, and *Rhizomucor miehei* (immobilized on epoxy-functionalized silica), *Candida antarctica* Novozyme 435 (immobilized on acrylic resin beads), and *Thermomyces lanuginosus* (immobilized on silica gel) have emerged as important tools. The results obtained with the current immobilized lipases exhibit comparable conversion values (45-99%). Table 2 shows the immobilized lipases produced from different microbial strains acting in biodiesel production.

Another important point about the immobilization of enzymes is the possibility of obtaining reusable enzymes. According to (Babaki et al. 2015), the use of the immobilization technique enables easy handling, recovery and recycling of the biocatalyst and hence lowers the cost. Immobilized lipases obtained from *Candida antarctica*, *Thermomyces lanuginosus*, and *Rhizomucor miehei* on epoxy-functionalized silica were quite stable and can be reused for 16 cycles without significant loss in activity (15%, respectively) (Babaki et al. 2015). Aybastıer and Demir 2014 studied operational stability of immobilized lipase on chitosan beads and reported that 74% of residual activity was found even after repeated use for five consecutive batches of 24 hours each (Aybastıer and Demir 2014).

3. Structured Lipids

3.1. General Features and Structure

Lipids are essential cellular components with a broad range of roles. These molecules of fatty acid and glycerol are soluble in organic solvents and their importance can be demonstrated by the number of biological and technological processes in which they are involved and the diseases they can cause in abnormal conditions. Lipids are source of energy, form biological membranes and contribute to transport of fat-soluble vitamins. They regulate gene expression, supply a favorable hydrophobic ambient to proteins, participate in the post-translational modification of proteins by the adding of acyl group and give rise to secondary messengers that act in cell signaling (Shahidi et al. 1997; Han and Gross 2005; Wenk 2005; Heidor et al. 2015). Besides, lipids can be modified chemically or enzymatically to acquire desirable properties that are employed by pharmaceutical, chemical and food industries and used for medical and nutritional applications in the form of added-value products (Hellner et al. 2010; Martin et al. 2010; Kim and Akoh 2015; Li et al. 2015).

These modified lipids can be triacylglycerols (TAGs) named structured lipids (SLs) and, by definition, SLs are fatty acids containing variable chain length attached to the glycerol backbone (Yang et al. 2005; Hellner et al. 2010; Kim and Akoh 2015). This name was coined for the first time in the review written by Babayan about medium chain tryacylglicerols (see below) in 1987 (Lee and Akoh 1998). TAGs consist of a glycerol backbone esterified to three fatty acids (Fox and Stachowiak 2007) and these fatty acids can be modified in their positional distribution and composition to obtain SLs (Martin et al. 2010).

According to the number of carbon (C) elements of fatty acid chains, SLs might have short (C_1-C_4) (SCFAs), medium (C_6-C_{12}) (MCFAs) or long (>C_{14}) (LCFAs) chain (Hellner et al. 2010). The first two SLs, SCFAs and MCFAs, have more solubility and rapid absorption than LCFAs. LCFAs have low absorption, are metabolized slowly and they deposit mainly in the human adipose tissue (Kim and Akoh 2015). Long chain fatty acids might be divided in saturated (SFAs) and polyunsaturated (PUFAs), according to the presence of double bonds. SFAs are interesting for production of low-calorie SLs and PUFAs have also received attention because of their health properties and food applications (Takahata et al. 1998; Martin et al. 2010; Bazinet and Layé 2014; Nakamura et al. 2014).

Vegetable oils and fats are mainly composed of tryacylglicerols (TAGs) and are among the most renewable feedstock of the chemical industry (Firdaus et al. 2014; Kadhum and Shamma 2015). Palm oil is considered the most important vegetable oil in the world followed just by soybean oil (Mba et al. 2015). Palm oil is responsible for 30% of the world's vegetable oil production (Sun et al. 2015). It is extracted from the palm tree and its major producers are Malaysia and Indonesia (Mancini et al. 2015). It contains mostly palmitic acid (16:0) (44%) and oleic acid (18:1) (39.2%) in the composition and despite the palm oil being used in the food industry; it has been the subject of studies evaluating the unhealthy effects of palm oil in the diet, mainly because palmitic acid increases serum cholesterol (Mancini et al. 2015; Odia et al. 2015).

Composition and positional changes in fatty acids are well known, such as the adding of new fatty acids or alteration of their position in glycerol structure that yield novel lipids with different physical characteristics from the original lipid, like melting point and thermal behavior (Martin et al. 2010; Kadhum and Shamma 2015; Kim and Akoh 2015). Thus the possibility of transforming a lipid in another form with desired attributes has received a crescent attention (Martin et al. 2010; Madeira Junior et al. 2012; Kim and Akoh 2015). In food and pharmaceutical industries, for instance, there are many available commercial products made from SLs with different purposes, such as substitutes of the infant milk fat ("Betapol," Loders Croklann Company; "Infant," Enzymotec USA Inc.), for chocolate ("Bohenin," Fuji Oil Company Ltd., Osaka) and cocoa butter production ("Caprenin," Procter and Gamble, USA) (Martin et al. 2010).

3.2. Structured Lipids Production

SLs might be produced through chemical or enzymatic process. There are pros and cons for these two processes, but the main difference consists in the specificity of the reaction: in chemical reaction, the distribution of fatty acids in the SL is random while in the enzymatic reaction is specific. The chemical process is more complex than synthesis mediated by enzymes. It contains four steps: pretreatment of the oils, addition of the catalyst, reaction and inactivation of the catalyst. In the reaction step, the acyl groups are hydrolyzed from fatty acids of interest and then they suffer a random reesterification on the glycerol, giving rise to SLs. Sodium methylate is one of the most used catalyst for esterification, but there are also others alkali metal alkylates utilized. The chemical process is less expensive and easier to scale up than enzymatic process, however the undesirable products formed in the whole process are difficulty to remove, many purification steps being necessary (Willis and Marangoni 1999; Yang et al. 2005; Martin et al. 2010).

On the other hand, enzymatically-catalyzed synthesis is less energy consuming, the final products are easier to purify and it is possible control which position the new fatty acid will be attached to the glycerol backbone. This process is catalyzed by two types of lipases: sn-1,3-specific lipases and nonspecific lipases (Hou and Shimada 2009). The first type is able to cleave at the sn-1 and sn-3 positions of TAGs and replace the original fatty acid with the desired one. However, the fatty acid bound at the sn-2 position on glycerol backbone is not modified by these lipases (Yang et al. 2005; Hou and Shimada 2009). These specific lipases are produced naturally by many microorganisms, such as *Aspergillus niger, Mucor miehei, Penicillium* spp. and *Candida* spp, and might be easily recovered from them (Hasan et al. 2009; Hou and Shimada 2009; Martin et al. 2010). The nonspecific lipases recognize all ester bonds of TAGs and are found in *C. rugosa, G. candidum, Pseudomonas* and *Burkholderia*. These enzymes can hydrolyze bonds at the 1,3- and 2-positions of TAG, but others hydrolyze 1,3-position more readily than 2-position (Hou and Shimada 2009).

Regarding the enzymatic reaction, SLs can be produced by direct esterification of the desired fatty acids onto glycerol or interesterification. Interesterification is the most used reaction and it includes acidolysis, alcoholysis, glycerolysis and transesterification (interchange of a acyl group from TAG for a acid fatty, an alcohol, a glycerol and esters, respectively) (Martin et al. 2010).

The main limitations of lipases-catalyzed reactions are the high cost of the enzymes used and the need of optimizing the conditions for each desirable SL because these enzymatic reactions are sensitive to water activity, temperature and pH (Willis and Marangoni 1999; Yang et al. 2005; Hellner et al. 2010; Martin et al. 2010).

3.3. Practical Applications of Structured Lipids

SLs and lipases are employed widely in food, chemical and pharmaceutical industries. In food industry, for example, SLs have been used as a milk fat substitute in infant formulas when there is not enough maternal milk or the quality is not suitable for breastfeeding (Li et al. 2015). The human milk fat is mainly made up of palmitic acid, a fatty acid that is attached at the sn-2 position of glycerol backbone (Xu 2000). On the other hand, palmitic acid at the sn-1,3 positions is naturally present in cow's milk and vegetable oils. However, this positional configuration leads to lower absorption of palmitic acid and calcium by the infants, because this fatty acid forms an insoluble compound with calcium ions in the intestines. Thus, the vegetable TAGs containing palmitic acid must be modified in order to obtain a healthier milk fat substitute, such as the commercial SL Betapol® (Loders Croklaan, USA). This product is already available and it was produced from unsaturated fatty acids and using tripalmitin (a sn-1,3 especific lipase) as enzymatic catalyst (Osborn and Akoh 2002; Li et al. 2015).

SLs are also used for margarine and butter production. These products are called "plastic" fats because they must be solid in refrigerators and spread when consumed. For this purpose, SCFAs and MCFAs are incorporated in the chemical process for production of TAGs with good temperature stability and spreadability (Osborn and Akoh 2002). In this context, many

SLs were chemically and enzymatically produced, for example palm stearin-sunflower oil (Lai et al. 1998) and butterfat-canola oil (Rousseau et al. 1996).

In the Osborn and Akoh review about SLs, many different studies trying to develop new SLs for enteral (oral feeding) and parenteral (intravenous feeding) nutrition using diverse combinations of fatty acids are listed. These structured lipids have differences in absorption and processing compared to their natural forms and it has been shown they have beneficial effects for the human health. It was demonstrated SLs improve immune response, decrease some cancer risk and can lower bloodstream cholesterol (Osborn and Akoh 2002; Sun et al. 2015).

4. Minor Hydrophobic Compounds

4.1. Sterol

The biotransformation of sterol has been studied for more than 60 years in reactions such as degradation, hydroxylation and redox modifications by microbial whole cells (Donova and Egorova 2012). One of the most successful industrial example of biotransformation is the production of steroids drugs and hormones (Fernandes et al. 2003). Another example is the conversion of β-sitosterol into 4-Androsteno-3,14-dione by *Mycobacterium vaccae* (Rumijowska et al. 1997) and into 4-androstene-3,17-dione (AD) and 1,4-androstenediene-3,17-dione (ADD) by *Mycobacterium* sp. (Sripalakit et al. 2006). Androstenedione (AD) a pro-hormone substance and key precursor of pharmaceutical sterol (Egorova et al. 2002).

4.2. Tocopherol

Tocopherols are lipid compounds and natural source of vitamin found in vegetable oil. It is a mixture of alpha, beta, gamma and delta tocopherol with fat-soluble antioxidant properties (Shimoda et al. 2007).

Vitamin E is an essential nutrient composed by tocopherols and tocotrienols and has a potential use in drugs, for its anticarcinogenesis, anti-aging, anticancer and anti-atherosclerosis properties. However, this vitamin is a lipophilic compound with low oral absorption. For this reason, the glycosylation of vitamin E appears as an alternative reaction in which converts Vitamin E into water-soluble and stable improving its bioavailability and pharmacological properties. Such reaction naturally occurs in some bacteria, as *Klebsiela pneumoniae*, *Xanthomonas campestris* and *Lactobacillus delbrueckii* (Shimoda et al. 2009). Other authors reported glycosylation of α- and δ-tocopherols by whole microorganism and enzyme, *Klebsiella pneumoniae* and cyclodextrin glucanotransferase (CGTase), respectively, into the corresponding β-glucosides and β-malto-oligosaccharides as tocopherol derivatives, These compounds present anti-allergic activity and potential to be applied as food additives (Shimoda et al. 2009).

Another biotransformation is the conversion of Vitamin E into Vitamin E acetate by acetylation. It is added in food and drugs due to its high stability in the presence of light and oxygen, and different physical properties such as solubility and miscibility. Vitamin E is generally administered as a drug in the form of all-rac-α-tocopheryl acetate (vitamin E acetate) or all-rac-α-tocopheryl succinate (vitamin E succinate) and is also added in food to increase vitamin E content with shelf life greater than the natural source of tocopherol (Torres et al. 2008).

4.3. Carotenoids

Carotenoids are lipids with anti-oxidative properties, applied mainly in the food industry and as agent for prevention of atherosclerosis and cancer (Schörken and Kempers 2009). Another application is based on its cleavage by several fungi and yeast or enzyme as lipoxygenases and fungal peroxidase into β-ionone, a floral, fruity and sweet aroma compound widely used in aroma and fragrance industries (Zorn et al. 2003; Nacke et al. 2012). The advantage of the bioprocess of β–ionone production is the increasing preference of the consumers for natural or bio or organic products, with market value approximately 10–100 times higher than the synthetic form (Nacke et al. 2012).

4.4. Phospholipids

Phospholipids (PLs) are amphiphilic lipids with application in the food industry as stabilizers, emulsifiers and antioxidants. Its conversion into structured phospholipids by interesterification with triacylglycerols using an immobilized lipase increases heat stability and emulsifying properties. Structured PLs can be used in emulsion systems or in coatings for different uses (Chmiel et al. 1999; Hama et al. 2015; Guo et al. 2005).

5. Metabolic Engineering of Biosynthesis Pathways of Vegetable Oils

The steady growth in demand for fuels and by-products arising from crude oil brings to the fore the need to seek new energy sources and raw materials that are not as harmful to the environment, since the concentration of compounds such as carbon dioxide in the atmosphere brings with it a wide range of issues related to climate change. In this scenario, the reduction in the use of compounds based on petroleum becomes necessary. Vegetable oils arise as a viable alternative to the use of these compounds, because their production is more sustainable since it is based on the utilization of the plant's metabolic machinery for such (Dyer and Mullen 2008). This is possible since the fatty acids originating from vegetable oils are structurally similar to hydrocarbons found in the crude oil, making them potentially environmentally friendly substituents (Dyer and Mullen 2008; Lee et al. 2015).

Currently, there are many known vegetable oils that are used as industrial raw materials and for human consumption, and most of them originate from four main crops: sunflowers, soybean, canola and palm. As a raw material for the food industry, vegetable oils are used for the production of many products such as margarine and oils and they are generally composed of fatty acids known as "common," for example, oleic and palmitic acids. Furthermore, there are different fatty acids that can be used by the chemical industry for the production of lubricants, soaps and plastics. These fatty acids, on the other hand, are called "uncommon" due to the fact that they have a differentiated structure that can be used for many industrial purposes (Lee et al. 2015). The composition and abundance of fatty acids in vegetable oils vary between plant species showing that there are genetic differences between them that favor, or do not favor, the production and accumulation of specific compounds. As the cultivars mentioned above were studied exhaustively over the years to be aimed at producing oils for human consumption, the metabolic pathways responsible for the oils of chemical industry were mitigated showing that is important to understand the metabolic biosynthesis pathways of these other oils so that it can be manipulated to favor the production of them. There is a great interest from the food industry for producing plants which oil composition is compatible to their patterns of production and sale, such as a plant which seed has oil composed mostly of what are known as healthy fatty acids, that is, of unsaturated chains. Similarly, industries that produce lubricants, paints, adhesives and plastics have also begun to show interest in the modification of vegetable oil composition in order to promote their specific production processes.

Therefore, an understanding of the metabolic pathways of biosynthesis pertaining to the vegetable oils and enzymes involved in this process is important so that it can be possible to maximize, improve and diversify the production of the same, reducing our dependence on petroleum and creating a more bio-based economy (Dyer and Mullen 2008). It is precisely in this context that metabolic engineering aims to act in this question: a) improving the metabolic biosynthesis pathways of oils of interest; b) introducing these pathways into other plants in order to facilitate the production of these oils; c) changing the composition of fatty acids in vegetable oils, to better adapt them to their purpose, whether for the food or chemical industry.

One example of metabolic engineering use by the food industry is the facilitation of the solidification of vegetable oils without the necessity of hydrogenation. This reaction is applied into the oils to undo the double bounds of the carbon backbone by adding a pair of atomic hydrogens into it. The problem of this reaction is the formation of isomers trans-unsaturated that are extremely harmful to the human health and are not present in natural vegetable oils. Liu et al. 2000 noticed that the cotton seed oil is rich in polyunsaturated fatty acids therefore there is a need to hydrogenate it for the production of margarines and other products with a more stable structure to keep viable for more time. This step led to the production of those trans-unsaturated fatty acids mentioned above. To resolve this problem, the researchers created a transgenic cotton with downregulation of the gene encoding the enzyme Stearoyl-ACP *D*-9 desaturase resulting in a plant with an increase of 38% in the production of the stearic acid, a saturated fatty acid with a 18 carbon atoms chain but neutral in the cholesterol and LDL blood levels (Kris-Etherton et al. 2005) allowing the production of margarine without the necessity of hydrogenating the vegetable oil. Beside stearic acid, all other saturated fatty acids should be avoided in the human diet due to their physical and chemical properties which led to many health problems. Regarding this issue, Kinney et al. 1996 created a modified soybean plant with the downregulation of the gene encoding the enzyme Oleoyl-*D*-12 desaturase resulting an increased proportion of oleic acid (unsaturated fatty acid) in comparison with the amount of saturated fatty acids in the soybean oil composition. This modification benefits the use of this oil for human consumption and makes it more stable in higher temperatures such as those encountered in frying.

Recently, Bhattacharya et al. 2015 knowing that the rate of undesirable saturated fatty acids as erucic acid in the *Brassica juncea* seed oil are much higher than the amount of stearic acid (a saturated fatty acid mentioned above, neutral in the balance of the cholesterol in the blood and the levels of LDL), constructed a transgenic plant of *Brassica juncea* with the carbon flux manipulated favoring the production of the healthy fatty acid over the erucic acid. This was made by the introduction of the gene *MlFatB* encoding the enzyme FatB thioesterase from *Madhuca longifolia* in the *B. juncea* plant. Analyzing the content of lipids in the seeds from the transgenic lines and the control lines (not transformed with the gene *MlFatB*). Bhattacharya et al. 2015 noticed a reduction of 71% in the content of erucic acid in the transgenic seeds and a considerable increase of the desirable stearic acid. Among other examples, one recent study led to the construction of a transgenic *Camelia sativa* and *Glycine max* (soybean) plants with the higher amount of the unusual lipids acetyl-triacylglycerols, in comparison with all other transgenic plants designed for this purpose of unusual fatty acid production. This was made by the downregulation by RNAi suppression of the enzyme diglyceride acyltransferase (DGAT1) which catalyzes the reaction of common triacylglycerols from acetyl-CoA and diacylglycerol and the heterologous expression of the gene encoding the diacylglycerol acetyltransferase (EaDAcT) from *Euonymus alatus*. This unusual lipid produced by metabolic engineering could be used for many industrial applications because its physical and chemical properties allow it. For example, its kinematic viscosity is in the range for diesel utilization without the need of transesterification, acetyl-triacylglycerols are less caloric than the common tryacylglicerols found in vegetable oils and it is possible to put these genetic modifications into oilseed crops such as soybean for human consumption, as was made in this study too, showing an increased production of the acetyl-triacylglycerols (Liu et al. 2015).

The industrial need for more sustainable, ecological and healthier feedstock is accompanied by the development of new biotechnological strategies to achieve this target. One of the approaches is the metabolic engineering of plant's pathways for the production of vegetable oils that can replace the environmental harmful feedstocks and be used as healthier foods. In this context, many efforts have successfully been made in the modification of the fatty acids composition of the vegetable oils to benefit the human consumption, with healthier lipids, or to benefit those lipids that are important for the chemical industry for the production of waxes, lubricants, plastics and fuels. These objectives can only be totally achieved if we expand our knowledge of the biosynthesis pathways of the oils and the genes involved in such process. Once it is accomplished, amazing results rises with new plants producing better compositions of lipids and producing lipids which can be used to replace harmful feedstocks arising from petroleum.

Conclusion

Vegetable oils are very important feedstock that can be used in many industrial processes. Although it is mainly used for basic nutrition, several vegetables oils present many functional properties, including antioxidant properties. However, with specific modification of vegetable oils many industrially relevant compounds can be obtained. Among the modified oils, there are structured lipids for food and pharmaceutical industries and biodiesel for fuel industry. In addition, many minor hydrophobic compounds present in vegetables can be modified for industrial interest. In this context, although traditional chemical methods can be used, the utilization of efficient biotechnological techniques, including biocatalysis and biotransformation, present many advantages, such as higher selectivity for the substrate, what does not cause the formation of toxic byproducts and the fact that enzymes and cells are considered environmentally friendly. In addition to enzymatic techniques, the genetic engineering of plants can be a future alternative to obtain desirable oils with specific properties. However, the necessity of better enzymes and new microorganisms with higher catalytic properties, added to new metabolic engineering technologies is highly necessary to the real day by day industrial production of modified vegetable oils.

REFERENCES

Abbaszaadeh, Ahmad, Ghobadian, Barat, Omidkhah, Mohammad Reza, and Gholamhassan Najafi. 2012. "Current Biodiesel Production Technologies: A Comparative Review." *Energy Conversion and Management* 63: 138–48. http://linkinghub.elsevier.com/retrieve/pii/S0196890412001185.

Abdelmoez, Wael, and Ahmad Mustafa. 2014. "Oleochemical Industry Future through Biotechnology." *Journal of oleo science* 63: 545–54. http://www. ncbi.nlm.nih.gov/pubmed/24881769.

Abraham, W. R., Hoffmann, H. M. R., Kieslich, K., Reng, G., Stumpf, B. 1986. "Microbial transformations of some monoterpenoids and sesquiterpenoids." In Enzymes in organic chemistry, edited by Ruth Porter and Sarah Clark. Ciba Foundation Symposium III Pitman, London, 146-160.

Agarwal, Avinash Kumar. 2007. "Biofuels (alcohols and Biodiesel) Applications as Fuels for Internal Combustion Engines." *Progress in Energy and Combustion Science* 33: 233–71. doi: 10.1016/j.pecs.2006.08.003.

Ahmad, A. L., N. H. Mat Yasin, C. J. C. Derek, and J. K. Lim. 2011. "Microalgae as a Sustainable Energy Source for Biodiesel Production: A Review." *Renewable and Sustainable Energy Reviews* 15: 584–93. http://linkinghub.elsevier.com/retrieve/pii/S1364032110003059.

Ashraful, A. M., H. H. Masjuki, M. A. Kalam, I. M. Rizwanul Fattah, S. Imtenan, S. A. Shahir, and H. M. Mobarak. 2014. "Production and Comparison of Fuel Properties, Engine Performance, and Emission Characteristics of Biodiesel from Various Non-Edible Vegetable Oils: A Review." *Energy Conversion and Management* 80: 202–28. http://linkinghub.elsevier.com/retrieve/pii/S019689041400082X.

Aybastıer, Ö., and C. Demir. 2014. "Immobilization of Candida Antarctica Lipase A on Chitosan Beads for the Production of Fatty Acid Methyl Ester from Waste Frying Oil." *Energy Sources, Part A: Recovery, Utilization, and Environmental Effects* 36(21): 2313–19. http://www.tandfonline. com/doi/abs/10.1080/15567036.2011.567233.

Babaki, Mohadese, Maryam Yousefi, Zohreh Habibi, Jesper Brask, and Mehdi Mohammadi. 2015. "Preparation of Highly Reusable Biocatalysts by Immobilization of Lipases on Epoxy-Functionalized Silica for Production of Biodiesel from Canola Oil." *Biochemical*

Engineering Journal 101: 23–31. http://www.sciencedirect.com/science/article/pii/S1369703X1500162 X.

Babayan, V K. 1987. "Medium Chain Triglycerides and Structured Lipids." *Lipids* 22(6): 417–20. http://www.ncbi.nlm.nih.gov/pubmed/3112486.

Bajaj, Akhil, Purva Lohan, Prabhat N. Jha, and Rajesh Mehrotra. 2010. "Biodiesel Production through Lipase Catalyzed Transesterification: An Overview." *Journal of Molecular Catalysis B: Enzymatic* 62(1): 9–14. http://linkinghub.elsevier.com/retrieve/pii/S1381117709002525.

Basri, Mahiran, Mohd Azlan Kassim, Rosfarizan Mohamad, and Arbakariya B. Ariff. 2013. "Optimization and Kinetic Study on the Synthesis of Palm Oil Ester Using Lipozyme TL IM." *Journal of Molecular Catalysis B: Enzymatic* 85-86: 214–19. http://www.sciencedirect.com/science/article/pii/S1381117712002652.

Bazinet, Richard P., and Sophie Layé. 2014. "Polyunsaturated Fatty Acids and Their Metabolites in Brain Function and Disease." *Nature reviews. Neuroscience* 15(12): 771–85. http://dx.doi.org/10.1038/nrn3820.

Bhattacharya, S., Sinha, S., Das, N., Maiti, M.K. 2015. "Increasing the stearate content in seed oil of Brassica juncea byheterologous expression of MlFatB affects lipid content and germination frequency of transgenic seeds." *Plant Physiology and Biochemistry* 96: 345-355. Acessed December 21, 2015. doi: 10.1016/j.plaphy.2015.08.015.

Biermann, Ursula, Bornscheuer, Uwe, Meier, Michael A.R., Metzger, Jürgen O., and Hans J. Schäfer. 2011. "Oils and Fats as Renewable Raw Materials in Chemistry." *Angewandte Chemie International Edition* 50: 3854–71. DOI: 10.1002/anie.201002767.

Calero, Juan, Cristóbal Verdugo, Diego Luna, Enrique D Sancho, Carlos Luna, Alejandro Posadillo, Felipa M Bautista, and Antonio A Romero. 2014. "Selective Ethanolysis of Sunflower Oil with Lipozyme RM IM, an Immobilized Rhizomucor Miehei Lipase, to Obtain a Biodiesel-like Biofuel, Which Avoids Glycerol Production through the Monoglyceride Formation." *New biotechnology* 31(6): 596–601. http://www.sciencedirect.com/science/article/pii/S1871678414000211.

Certik, M. 2008. "Biotransformation of Oils to Value-added Compounds." In Biocatalysis and Bioenergy, edited by Ching T. Hou and Jei-Fu Shaw, 571-585. New Jersey: John Wiley and Sons Inc.

Cheong, Wei Fun, Markus R. Wenk, and Guanghou Shui. 2014. "Comprehensive Analysis of Lipid Composition in Crude Palm Oil Using Multiple Lipidomic Approaches." *Journal of Genetics and Genomics* 41(5): 293–304. doi:10.1016/j.jgg.2014.04.002.

Chmiel, Oliver, Nicholas Melachouris, and Helmut Traitler. 1999. "Process for the Interesterification of Phospholipids." Patent number US 5989599 A.

Christopher, Lew P, Hemanathan Kumar, and Vasudeo P Zambare. 2014. "Enzymatic Biodiesel : Challenges and Opportunities." *Applied Energy* 119: 497–520. doi: 10.1016/j.apenergy.2014.01.017.

Daiha, Karina de Godoy, Renata Angeli, Sabrina Dias de Oliveira, and Rodrigo Volcan Almeida. 2015. "Are Lipases Still Important Biocatalysts? A Study of Scientific Publications and Patents for Technological Forecasting." *Plos One* 10(6): e0131624. http://dx.plos.org/10.1371/journal.pone.0131624.

Demirbas, Ayhan. 2009. "Progress and Recent Trends in Biodiesel Fuels." *Energy Conversion and Management* 50(1): 14–34. http://linkinghub. elsevier.com/retrieve/pii/S0196890408003294.

Donova, Marina V., and Olga V. Egorova. 2012. "Microbial Steroid Transformations: Current State and Prospects." *Applied Microbiology and Biotechnology* 94(6): 1423–47. doi: 10.1007/s00253-012-4078-0. doi: 10.1007/s00253-012-4078-0.

Dyer, J. M., Mullen, R. T. 2008. "Engineering plant oils as high-value industrial feedstocks for biorefining: the need for underpinning cell biology research." *Physiologia Plantarum* 132: 11–22. Acessed December 11, 2015. doi: 10.1111/j.1399-3054.2007.01021.x.

Egorova, Olga V et al. 2002. "Production of Androstenedione Using Mutants ofMycobacterium Sp." *Journal of Chemical Technology & Biotechnology* 77(2): 141–47. doi: 10.1002/jctb.536.

Fernandes, P. et al. 2003. "Microbial Conversion of Steroid Compounds: Recent Developments." *Enzyme and Microbial Technology* 32(6): 688–705. doi: 10.1016/S0141-0229(03)00029-2.

Firdaus, Maulidan, Michael a R Meier, Ursula Biermann, and Jürgen O. Metzger. 2014. "Renewable Co-Polymers Derived from Castor Oil and Limonene." *European Journal of Lipid Science and Technology* 116: 31–36. DOI: 10.1002/ejlt.201300206.

Fox, N. J., and G. W. Stachowiak. 2007. "Vegetable Oil-Based lubricants—A Review of Oxidation." *Tribology International* 40(7): 1035–46. http://linkinghub.elsevier.com/retrieve/pii/S0301679X06003185.

Guo, Zheng, Anders F. Vikbjerg, and Xuebing Xu. 2005. "Enzymatic Modification of Phospholipids for Functional Applications and Human Nutrition." *Biotechnology Advances* 23(3): 203–59. doi:10.1016/j. biotechadv.2005.02.001.

Haas, Michael J. 2005. "Improving the Economics of Biodiesel Production through the Use of Low Value Lipids as Feedstocks: Vegetable Oil Soapstock." *Fuel Processing Technology* 86: 1087–96. doi: 10.1016/j.fuproc.2004.11.004.

Hama, Shinji, Chiaki Ogino, and Akihiko Kondo. 2015. "Enzymatic Synthesis and Modification of Structured Phospholipids: Recent Advances in Enzyme Preparation and Biocatalytic Processes." *Applied Microbiology and Biotechnology* 99(19): 7879–91.

Han, Xianlin, and Richard W. Gross. 2005. "Shotgun Lipidomics: Electrospray Ionization Mass Spectrometric Analysis and Quantitation of Cellular Lipidomes Directly from Crude Extracts of Biological Samples." *Mass Spectrometry Reviews* 24(3): 367–412. http://doi.wiley.com/10.1002/mas.20023.

Hasan, Fariha, Aamer Ali Shah, and Abdul Hameed. 2009. "Methods for Detection and Characterization of Lipases: A Comprehensive Review." *Biotechnology advances* 27(6): 782–98. doi:10.1016/j.biotechadv. 2009.06.001.

Heidor, Renato, de Conti, Aline, Ortega, Juliana F., Furtado, Kelly S., Silva, Roberta C., Tavares, Paulo E. L. M., Purgatto, Eduardo, Ract, Juliana N.R., de Paiva, Sérgio A. R., Gioielli, Luiz A., Pogribny, Igor P., and Fernando S. Moreno. 2015. "The Chemopreventive Activity of Butyrate-Containing Structured Lipids in Experimental Rat Hepatocarcinogenesis." *Molecular Nutrition & Food Research* 60(2):420-9. doi: 10.1002/mnfr. 201500643.

Hellner, Gabriella, Tőke, Enikő R., Nagy, Viviána, Szakács, György, and László Poppe. 2010. "Integrated Enzymatic Production of Specific Structured Lipid and Phytosterol Ester Compositions." *Process Biochemistry* 45: 1245–50. doi:10.1016/j.procbio.2010.04.008.

Hou, C. T., and Y. Shimada. 2009. “Lipases.” In *Encyclopedia of Microbiology*, ed. Moselio Schaechter. Elsevier, 385–92.

Janaun, Jidon, and Naoko Ellis. 2010. “Perspectives on Biodiesel as a Sustainable Fuel.” *Renewable and Sustainable Energy Reviews* 14(4): 1312–20. http://linkinghub.elsevier.com/retrieve/pii/S1364032109002913.

Kabbashi, Nassereldeen Ahmed, Nurudeen Ishola Mohammed, Md Zahangir Alam, and Mohamed Elwathig Saeed Mirghani. 2015. “Hydrolysis of Jatropha Curcas Oil for Biodiesel Synthesis Using Immobilized Candida Cylindracea Lipase.” *Journal of Molecular Catalysis B: Enzymatic* 116: 95–100. http://linkinghub.elsevier.com/retrieve/pii/S1381117715000661.

Kadhum, Abdul Amir H., and M. Najeeb Shamma. 2015. “Edible Lipids Modification Processes: A Review.” *Critical Reviews in Food Science and Nutrition* 5:0. http://dx.doi.org/10.1080/10408398.2013.848834.

Kaieda, Masaru, Taichi Samukawa, Akihiko Kondo, and Hideki Fukuda. 2001. “Effect of Methanol and Water Contents on Production of Biodiesel Fuel from Plant Oil Catalyzed by Various Lipases in a Solvent-Free System.” *Journal of Bioscience and Bioengineering* 91(1): 12–15. doi: 10.1016/S1389-1723(01)80103-1.

Kim, Byung Hee, and Casimir C. Akoh. 2015. “Recent Research Trends on the Enzymatic Synthesis of Structured Lipids.” *Journal of Food Science* 80(8): C1713–24. http://doi.wiley.com/10.1111/1750-3841.12953.

Kim, Sok Joong, Sang Min Jung, Yong Cheol Park, and Kyungmoon Park. 2007. “Lipase Catalyzed Transesterification of Soybean Oil Using Ethyl Acetate, an Alternative Acyl Acceptor.” *Biotechnology and Bioprocess Engineering* 12(4): 441–45.

Kinney, A. J. 1996. “Development of genetically engineered soybean oils for food applications.” *Journal of Food Lipids* 3: 273–292. doi: 10.1111/j.1745-4522.1996.tb00074.x.

Knothe, Gerhard. 2010. “Biodiesel and Renewable Diesel: A Comparison.” *Progress in Energy and Combustion Science* 36(3): 364–73. http://linkinghub.elsevier.com/retrieve/pii/S0360128509000677.

Köse, Öznur, Melek Tüter, and H. Ayşe Aksoy. 2002. “Immobilized Candida Antarctica Lipase-Catalyzed Alcoholysis of Cotton Seed Oil in a Solvent-Free Medium.” *Bioresource Technology* 83(2): 125–29. http://www.sciencedirect.com/science/article/pii/S0960852401002036.

Kris-Etherton P. M., Griel A. E., Psota T. L., Gebauer S. K., Zhang J., Etherton T. D. 2005. “Dietary stearic acid and risk of cardiovascular disease: intake, sources, digestion, and absorption.” *Lipids*. 40:1193–1200. Acessed December 21, 2015. doi: 10.1007/s11745-005-1485-y.

Kumar, Mukesh, and M. P. Sharma. 2015. “Assessment of Potential of Oils for Biodiesel Production.” *Renewable and Sustainable Energy Reviews* 44: 814–23. http://www.sciencedirect.com/science/article/pii/S136403211500 0234.

Kuo, Ting-Chun, Jei-Fu Shaw, and Guan-Chiun Lee. 2015. “Conversion of Crude Jatropha Curcas Seed Oil into Biodiesel Using Liquid Recombinant Candida Rugosa Lipase Isozymes.” *Bioresource Technology* 192: 54–59. http://www.sciencedirect.com/science/article/pii/S0960852415006720.

Lai, O.M., H.M. Ghazali, and C. L. Chong. 1998. "Effect of Enzymatic Transesterification on the Melting Points of Palm Stearin-Sunflower Oil Mixtures." *Journal of the American Oil Chemists' Society* 75(7): 881–86. http://link.springer.com/10.1007/s11746-998-0241-2.

Lapuerta, M, O Armas, and J Rodriguez fernandez. 2008. "Effect of Biodiesel Fuels on Diesel Engine Emissions." Progress in Energy and Combustion Science 34(2): 198–223. http://linkinghub.elsevier.com/retrieve/pii/S0360128507000421.

Lee, K. R., Chen, G. Q., Kim H. U. 2015. "Current progress towards the metabolic engineering of plant seed oil for hydroxy fatty acids production." *Plant Cell Reports* 34:603–615. Acessed December 2, 2015. doi: 10.1007/s00299-015-1736-6.

Lee, Ki-Teak, and Casimir C. Akoh. 1998. "Structured Lipids: Synthesis and Applications." *Food reviews international* 14(1): 17–34. doi: 10.1080 /87559129809541148.

Li, Changmo, Yunping Yao, Guozhong Zhao, Wen Cheng, Huilin Liu, Chunyang Liu, Zhen Shi, Yao Chen, and Shuo Wang. 2011. "Comparison and Analysis of Fatty Acids, Sterols, and Tocopherols in Eight Vegetable Oils." *Journal of Agricultural and Food Chemistry* 59(23): 12493–98. doi: 10.1021/jf203760k.

Li, Ruoyu, Jamal S. M. Sabir, Nabih A. Baeshen, and Casimir C. Akoh. 2015. "Enzymatic Synthesis of Refined Olive Oil-Based Structured Lipid Containing Omega -3 and -6 Fatty Acids for Potential Application in Infant Formula." *Journal of food science* 80(11): H2578–84. http://www.ncbi.nlm.nih.gov/pubmed/26408984.

Li, Xin, and Hui Ying Yu. 2014. "Characterization of an Organic Solvent-Tolerant Lipase from Haloarcula Sp. G41 and Its Application for Biodiesel Production." *Folia Microbiologica*: 455–63. doi: 10.1007/s12223-014-0320-8.

Liu, J., Rice, A., McGlew, K., Shaw, V., Park, H., Clemente, T., Pollard, M. Ohlrogge, J. B., Durrett, T. P. 2015. "Metabolic engineering of oilseed crops to produce high levels of novel acetyl glyceride oils with reduced viscosity,freezing point and calorific value." *Plant Biotechnology Journal* 13: 858–865. Acessed December 12, 2015. Acessed December 1, 2015. doi: 10.1111/pbi.12325.

Liu, Q., Singh, S., Green, A. 2000. "Genetic modification of cotton seed oil using inverted-repeat gene-silencing techniques." *Biochemical Society Transactions* 28: 927-929. Acessed December 2, 2015. doi: 10.1042/0300-5127:0280927.

Lu, Jike, Yawei Chen, Fang Wang, and Tianwei Tan. 2009. "Effect of Water on Methanolysis of Glycerol Trioleate Catalyzed by Immobilized Lipase Candida Sp. 99 – 125 in Organic Solvent System." Journal of Molecular Catalysis 56(2-3): 122–25. http://linkinghub.elsevier.com/retrieve/pii/S1381117708001215.

Madeira Junior, José V., Contesini, Fabiano J., Speranza, Paula, Sato, Hélia H., Macedo, Gabriela A., and Patrícia de Oliveira Carvalho. 2012. "Microbial Lipases and Their Applications in Structured Lipids." In *Lipase: Functions, Synthesis and Role in Disease*, edited by Hamdi Sassi and Sofien Cannamela. Nova Publishers.

Mancini, Annamaria, Imperlini, Esther, Nigro, Ersilia, Montagnese, Concetta, Daniele, Aurora, Orrù, Stefania, and Pasqualina Buono. 2015. "Biological and Nutritional Properties of Palm Oil and Palmitic Acid: Effects on Health." *Molecules* 20(9): 17339–61. http://www.mdpi.com/1420-3049/20/9/17339/.

Martin, D., G. Reglero, and Francisco J. Señoráns. 2010. "Oxidative Stability of Structured Lipids." *European Food Research and Technology* 231(5): 635-653. http://link.springer.com/article/10.1007/s00217-010-1324-5.

Mba, Ogan I., Marie-Josée Dumont, and Michael Ngadi. 2015. "Palm Oil: Processing, Characterization and Utilization in the Food Industry – A Review." *Food Bioscience* 10: 26–41. http://www.sciencedirect.com/science/article/pii/S221242921500005X.

Meher, L, D Vidyasagar, and S Naik. 2006. "Technical Aspects of Biodiesel Production by Transesterification—a Review." *Renewable and Sustainable Energy Reviews* 10(3): 248–68. doi: 10.1016/j.rser.2004.09.002.

Modi, Mukesh Kumar, J. R. C. Reddy, B. V. S. K. Rao, and R. B. N. Prasad. 2007. "Lipase-Mediated Conversion of Vegetable Oils into Biodiesel Using Ethyl Acetate as Acyl Acceptor." *Bioresource Technology* 98(6): 1260–64. http://linkinghub.elsevier.com/retrieve/pii/S0960852406002008.

Mofijur, M., M. G. Rasul, and J. Hyde. 2015. "Recent Developments on Internal Combustion Engine Performance and Emissions Fuelled With Biodiesel-Diesel-Ethanol Blends." Procedia Engineering 105(Icte 2014): 658–64. http://linkinghub.elsevier.com/retrieve/pii/S1877705815008425.

Mohsin, R., Z. A. Majid, A. H. Shihnan, N. S. Nasri, and Z. Sharer. 2014. "Effect of Biodiesel Blends on Engine Performance and Exhaust Emission for Diesel Dual Fuel Engine." *Energy Conversion and Management* 88(x): 821–28. http://www.sciencedirect.com/science/article/pii/S019689041400 8280.

Molina, Gustavo, Mariana R. Pimentel, and Gláucia M. Pastore. 2013. "Pseudomonas: A Promising Biocatalyst for the Bioconversion of Terpenes." *Applied Microbiology and Biotechnology* 97(5): 1851–64. http://link.springer.com/10.1007/s00253-013-4701-8.

Mosarof, M. H., M. A. Kalam, H. H. Masjuki, A. M. Ashraful, M. M. Rashed, H. K. Imdadul, and I. M. Monirul. 2015. "Implementation of Palm Biodiesel Based on Economic Aspects, Performance, Emission, and Wear Characteristics." Energy Conversion and Management 105: 617–29. http://linkinghub.elsevier.com/retrieve/pii/S01968904 15007669.

Nacke, Christoph, Sonja Hüttmann, Maria M. W. Etschmann, and Jens Schrader. 2012. "Enzymatic Production and in Situ Separation of Natural β-Ionone from β-Carotene." *Journal of Industrial Microbiology & Biotechnology* 39(12): 1771–78. doi: 10.1007/s10295-012-1182-1.

Nakamura, Manabu T., Barbara E. Yudell, and Juan J. Loor. 2014. "Regulation of Energy Metabolism by Long-Chain Fatty Acids." *Progress in lipid research* 53: 124–44. http://www.sciencedirect.com/science/article/pii/S0163782713000696.

No, Soo-Young. 2011. "Inedible Vegetable Oils and Their Derivatives for Alternative Diesel Fuels in CI Engines: A Review." *Renewable and Sustainable Energy Reviews* 15(1): 131–49. http://linkinghub.elsevier. com/retrieve/pii/S1364032110002649.

Odia, Osaretin J., Sandra Ofori, and Omosivie Maduka. 2015. "Palm Oil and the Heart: A Review." *World Journal of Cardiology* 7(3): 144. doi: 10.4330/wjc.v7.i3.144.

Oliveira, A. C., and M. F. Rosa. 2006. "Enzymatic Transesterification of Sunflower Oil in an Aqueous-Oil Biphasic System." *J. Am. Oil Chem. Soc.* 83(1): 21–25. doi: 10.1007/s11746-006-1170-6.

Oliveira, D, I Filho, M Luccio, C Faccio, C Rosa, JP Bender, and Et. 2005. "Kinetics of Enzyme-Catalyzed Alcoholysis of Soybean Oil in N-Hexane." In *Twenty-Sixth Symposium on Biotechnology for Fuels and Chemicals*, 231–41. doi: 10.1385/ABAB:121:1-3:0231.

Ondul, E., Dizge, N., Keskinler, B., and Albayrak, N. 2015. "Biocatalytic Production of Biodiesel from Vegetable Oils" In *Biofuels - Status and Perspective*, edited by Krzysztof Biernat, 21-37. InTech.

Osborn, H. T., and C. C. Akoh. 2002. "Structured Lipids-Novel Fats with Medical, Nutraceutical, and Food Applications." *Comprehensive Reviews in Food Science and Food Safety* 1(3): 110–20. http://doi.wiley.com/10.1111/j.1541-4337.2002.tb00010.x.

Pal, Amit, Raj Kumar Singh, and Shashank Mohan. 2015. "Biodiesel Conversionion of High FFA Neem Oil by Blending It with Low FFA Sesame Oil." 4(3): 127–30. http://www.jsirjournal.com/Vol4_Issue3_04.pdf.

Pizarro, Ana V Lara, and Enoch Y. Park. 2003. "Lipase-Catalyzed Production of Biodiesel Fuel from Vegetable Oils Contained in Waste Activated Bleaching Earth." *Process Biochemistry* 38(7): 1077–82. doi: 10.1016/S0032-9592(02)00241-8.

Poppe, Jakeline K., Cristina Garcia-Galan, Carla R. Matte, Roberto Fernandez-Lafuente, Rafael C. Rodrigues, and Marco Antônio Z Ayub. 2013. "Optimization of Synthesis of Fatty Acid Methyl Esters Catalyzed by Lipase B from Candida Antarctica Immobilized on Hydrophobic Supports." *Journal of Molecular Catalysis B: Enzymatic* 94: 51–56.http://dx.doi.org/10.1016/j.molcatb.2013.05.010.

Poppe, Jakeline Kathiele, Roberto Fernandez-Lafuente, Rafael C. Rodrigues, and Marco Antônio Záchia Ayub. 2015. "Enzymatic Reactors for Biodiesel Synthesis: Present Status and Future Prospects." *Biotechnology Advances* 33(5): 511–25. http://linkinghub.elsevier.com/retrieve/pii/S0734975015000245.

Ribeiro, Bernardo Dias, Aline Machado de Castro, Maria Alice Zarur Coelho, and Denise Maria Guimarães Freire. 2011. "Production and Use of Lipases in Bioenergy: A Review from the Feedstocks to Biodiesel Production." *Enzyme Research* 2011: 1–16. http://www.hindawi.com/journals/er/2011/615803/.

Robles-Medina, A., P.A. González-Moreno, L. Esteban-Cerdán, and E. Molina-Grima. 2009. "Biocatalysis: Towards Ever Greener Biodiesel Production." Biotechnology Advances 27(4): 398–408. http://linkinghub. elsevier.com/retrieve/pii/S0734975009000445.

Rousseau, Dérick, Arthur R. Hill, and Alejandro G. Marangoni. 1996. "Restructuring Butterfat through Blending and Chemical Interesterification. 3. Rheology." *Journal of the American Oil Chemists' Society* 73(8): 983–89. http://link.springer.com/10.1007/BF02523405.

Royon, D., M. Daz, G. Ellenrieder, and S. Locatelli. 2007. "Enzymatic Production of Biodiesel from Cotton Seed Oil Using T-Butanol as a Solvent." *Bioresource Technology* 98(3): 648–53. doi: 10.1016/j.biortech.2006.02.021.

Rumijowska, A., K. Lisowska, and L. Sedlaczek. 1997. "Transformation of Sterols by Mycobacterium vaccae: Effect of Lecithin on the Permeability of Cell Envelopes to Sterols." *World Journal of Microbiology and Biotechnology* 13(1):89–95. doi: 10.1007/BF02770813.

Sagiroglu, Ayten. 2008. "Conversion of Sunflower Oil to Biodiesel by Alcoholysis Using Immobilized Lipase." *Artificial cells, blood substitutes, and immobilization biotechnology* 36(2): 138–49. doi: 10.1080/10731190801932124.

Sambanthamurthi, R., Kalyana Sundram, and Yew-Ai Tan. 2000. "Chemistry and biochemistry of palm oil." *Progress in Lipid Research* 39(6): 507-558. doi:10.1016/S0163-7827(00)00015-1.

Schörken, Ulrich, and Peter Kempers. 2009. "Lipid Biotechnology: Industrially Relevant Production Processes." *European Journal of Lipid Science and Technology* 111(7): 627–45. http://doi.wiley.com/10.1002/ejlt.200900057.

Shah, Shweta, Shweta Sharma, and M. N. Gupta. 2004. "Biodiesel Preparation by Lipase-Catalyzed Transesterification of Jatropha Oil." *Energy and Fuels* 18(1): 154–59. doi: 10.1021/ef030075z.

Shahidi, F., P. K. J. P. D. Wanasundara, and U. N. Wanasundara. 1997. "Changes in edible fats and oils during processing." *Journal of Food Lipids* 4(3): 199–231. http://doi.wiley.com/10.1111/j.1745-4522.1997. tb00093.x.

Sharma, Shivika, and Shamsher S. Kanwar. 2014. "Organic Solvent Tolerant Lipases and Applications." *The Scientific World Journal* 2014. 15 pages. 10.1155/2014/625258.

Sharma, Y. C., B. Singh, and S. N. Upadhyay. 2008. "Advancements in Development and Characterization of Biodiesel: A Review." *Fuel* 87(12): 2355–73. http://linkinghub.elsevier.com/retrieve/pii/S0016236 10800029X.

Shimoda, Kei, Kei, Shimoda, Kondo Yoko, Akagi Masaaki, Abe Koichi, Hamada Hatsuyuki, and Hamada Hiroki. 2007. "Glycosylation of Tocopherols by Cultured Cells of Eucalyptus perriniana." *Phytochemistry* 68(21): 2678–83. doi:10.1016/j.phytochem.2007. 05.036.K.

Shimoda, Kei, Masaaki Akagi, and Hiroki Hamada. 2009. "Production of Beta-Maltooligosaccharides of Alpha- and Delta-Tocopherols by Klebsiella pneumoniae and Cyclodextrin Glucanotransferase as Anti-Allergic Agents." *Molecules* (Basel, Switzerland) 14(8): 3106–14. doi: 10.3390/molecules14083106.

Silitonga, A. S., H. H. Masjuki, T. M. I. Mahlia, Hwai Chyuan Ong, and W. T. Chong. 2013. "Experimental Study on Performance and Exhaust Emissions of a Diesel Engine Fuelled with Ceiba Pentandra Biodiesel Blends." *Energy Conversion and Management* 76: 828–36. http://linkinghub.elsevier.com/retrieve/pii/S0196890413004913.

Soler, Lorena, Andrés Illanes, and Lorena Wilson. 2016. "Immobilization of Alcaligenes Sp. Lipase as Catalyst for the Transesterification of Vegetable Oils to Produce Biodiesel." *Catalysis Today* 259: 177–82. http://linkinghub.elsevier.com/retrieve/pii/S0920586 115003922.

Sripalakit, Pattana, Uthai Wichai, and Aurasorn Saraphanchotiwitthaya. 2006. "Biotransformation of Various Natural Sterols to Androstenones by Mycobacterium Sp. and Some Steroid-Converting Microbial Strains." *Journal of Molecular Catalysis B: Enzymatic* 41(1-2): 49–54. doi: 10.1016/j.molcatb.2006.04.007.

Sun, Ye, Neelakantan N., Wu Y., Lote-Oke R., Pan A., and van Dam R. M. 2015. "Palm Oil Consumption Increases LDL Cholesterol Compared With Vegetable Oils Low in Saturated Fat in a Meta-Analysis of Clinical Trials." *The Journal of nutrition* 145(7): 1549-58. doi: 10.3945/jn. 115.210575.

Takahata, K., K. Monobe, M. Tada, and P. C. Weber. 1998. "The Benefits and Risks of N-3 Polyunsaturated Fatty Acids." *Bioscience, biotechnology, and biochemistry* 62(11): 2079–85. http://www.ncbi.nlm.nih.gov/pubmed/9972229.

Talukder, Md Mahabubur Rahman, Jin Chuan Wu, Thi Bich Van Nguyen, Ng Mei Fen, and Yeo Li Shi Melissa. 2009. "Novozym 435 for Production of Biodiesel from Unrefined Palm Oil: Comparison of Methanolysis Methods." *Journal of Molecular Catalysis B: Enzymatic* 60(3-4): 106–12. doi: 10.1016/j.molcatb.2009.04.004.

Tan, T., Lu, J., Nie, K., Zhang, H., Deng, L., Wang, F. 2010. "Progress on biodiesel production with enzymatic catalysis in China." *Chinese Journal of Biotechnology*. 26(7): 903-906.

Tan, Tianwei, Jike Lu, Kaili Nie, Li Deng, and Fang Wang. 2010. "Biodiesel Production with Immobilized Lipase : A Review." *Biotechnology Advances* 28(5): 628–34. doi: 10.1016/j.biotechadv.2010.05.012.

Teixeira, Camilo B., Madeira Junior, Jose V., and Gabriela A. Macedo. 2014. "Biocatalysis combined with physical technologies for development of a green biodiesel process." *Renewable and Sustainable Energy Reviews*. 33:333-343. doi: 10.1016/j.rser. 2014.01.072..

Torres, Pamela, Dolores Reyes-Duarte, Nieves López-Cortés, Manuel Ferrer, Antonio Ballesteros and Francisco J. Plou. "Acetylation of Vitamin E by Candida antarctica Lipase B Immobilized on Different Carriers." *Process Biochemistry* 43(2): 145–53. doi:10.1016/j.procbio.2007.11.008.

Vyas, Amish P., Jaswant L. Verma, and N. Subrahmanyam. 2010. "A Review on FAME Production Processes." *Fuel* 89(1): 1–9. http://linkinghub.elsevier.com/retrieve/pii/ S0016236109003755.

Wenk, Markus R. 2005. "The Emerging Field of Lipidomics." *Nature Reviews Drug Discovery* 4(7): 594–610. http://www.nature.com/doifinder/10.1038/nrd1776.

Willis, Wendy M., and Alejandro G. Marangoni. 1999. "Assessment of Lipase- and Chemically Catalyzed Lipid Modification Strategies for the Production of Structured Lipids." *Journal of the American Oil Chemists' Society* 76(4): 443–50. http://link.springer.com/10.1007/s11746-999-0022-6.

Xia, Sing, and Richard C. Larock. 2010. "Vegetable oil-based polymeric materials: synthesis, properties, and applications." *The Royal Society of Chemistry*. 12:1893–1909. Accessed December 20, 2015. doi:10.1039/c0gc00264.

Xie, Wenlei, and Jianlong Wang. 2012. "Immobilized Lipase on Magnetic Chitosan Microspheres for Transesterification of Soybean Oil." *Biomass and Bioenergy* 36: 373–80. http://dx.doi.org/10.1016/j.biombioe. 2011.11.006.

Xu, Xuebing. 2000. "Production of Specific-Structured Triacylglycerols by Lipase-Catalyzed Reactions: A Review." *European Journal of Lipid Science and Technology* 102(4): 287–303. doi: 10.1002/(sici)1438-9312(200004)102:4<287::aid-ejlt287>3.0.co;2-q.

Yang, T., M. Fruekilde, and X. Xu. 2005. "Suppression of Acyl Migration in Enzymatic Production of Structured Lipids through Temperature Programming." *Food Chemistry* 92(1): 101–7. http://www.sciencedirect. com/science/article/pii/S030881460400562X (December 15, 2015).

Zarei, Alireza, Nor Aishah Saidina Amin, Amin Talebian-Kiakalaieh, and Nor Azimah Mohd Zain. 2014. "Immobilized Lipase-Catalyzed Transesterification of Jatropha Curcas Oil: Optimization and Modeling." *Journal of the Taiwan Institute of Chemical Engineers* 45(2): 444–51. http://dx.doi.org/10.1016/j.jtice.2013.05.015.

Zhang, Kai-Pei, Jing-Qi Lai, Ze-Lin Huang, and Zhen Yang. 2011. "Penicillium Expansum Lipase-Catalyzed Production of Biodiesel in Ionic Liquids." *Bioresource technology* 102(3): 2767–72. http://dx.doi.org/10.1016/j.biortech.2010.11.057.

Zorn, H., S. Langhoff, M. Scheibner, and R. G. Berger. 2003. "Cleavage of Beta, beta-Carotene to Flavor Compounds by Fungi." *Applied microbiology and biotechnology* 62(4): 331–36. DOI: 10.1007/s00253-003-1309-4.

Zou, L, and S Atkinson. 2003. "Characterising Vehicle Emissions from the Burning of Biodiesel Made from Vegetable Oil." *Environ Technol* 24(10): 1253–60. http://www.ncbi.nlm.nih.gov/entrez/query.fcgi?cmd=Retrieve&db=PubMed&dopt=Citation&list _uids=14669805.

In: Vegetable Oil
Editor: Brittany Holt

ISBN: 978-1-63485-128-2

Chapter 2

ASSESSING AUTHENTICITY OF VEGETABLE OILS

Nicoleta-Aurelia Chira**, *Raluca Stan and Sorin Ioan Roşca
University POLITEHNICA of Bucharest, Romania
Faculty of Applied Chemistry and Materials Science
"C. Nenitescu" Department of Organic Chemistry, Bucharest, Romania

ABSTRACT

In recent years the production of vegetable oils (sunflower, linseed, rapeseed, soybean, olive, etc.) has significantly increased, oils being considered a major economic resource, mainly used in the food industry or as raw materials for the biodiesel production, as well as for the oleochemical and pharmaceutical industries.

Authentication of vegetable oils refers mainly to two aspects. On one hand, it is related to establishing conformity according to label - in terms of plant species, part of the plant (e.g., discrimination between palm oil and palm kernel oil), geographical origin, crop year, processing technique (differentiation of minimum processed oils such as cold pressed oils from refined oils). On the other hand, detection of adulteration of expensive vegetable oils and fats such as olive oil or cocoa butter with cheaper oils is another important issue concerning oils authentication.

In the last decade, researchers in the food quality control field, showed an increased interest to establish methods for products authentication. This trend results from the consumers enhanced demand for quality products. Vegetable oils authentication by means of establishing the geographical origin of oil seeds goes with the new trend of protecting the local food products. Protected designation of origin (PDO), protected geographical indication (PGI) and Traditional Specialty Guaranteed (TSG) are geographical indications defined in European Union Law to protect regional foods and to eliminate the dishonest practices. Those laws are inflicted in the European Community since 1992 and gradually extended to the non-UE countries by various joint agreements.

The fatty acid composition of vegetable oils is the main factor influencing their nutritional value and properties. In recent years, new and efficient methods for the determination of the fatty acids profile of different kind of samples (e.g., vegetable oils, lipid extracts from different vegetable or animal tissues) have evolved, being partially fed by the general trend in the chemical literature concerning the food authenticity

*Corresponding author: 1-7 Gh. Polizu Str., Sector 1, 011061 Bucharest, Romania. nicoleta.chira@chimie.upb.ro.

assessment. Among the most envisaged physical methods of analysis, vibrational spectroscopy (NIR, FT-IR and Raman), nuclear magnetic resonance (^{1}H-, ^{13}C-NMR) and chromatographic techniques (GC, HPLC) can provide structural and compositional information useful for vegetable oils authentication. Special attention has been focused on ^{1}H-NMR spectroscopy which leads to a global fatty acids profile (in terms of linolenic, linoleic, mono-unsaturated and saturated fatty acids) in a non-invasive manner and mild analysis conditions which do not require chemical transformations prior to/during analysis, as compared to chromatographic techniques. Moreover, spectra (both NMR or vibrational) can be considered as fingerprints of samples taken into analysis, thus being helpful for authentication purposes.

Authentication of a food product is almost impossible without the use of statistical methods applied to data bases obtained from genuine samples. Among these methods, multivariate statistical analysis - *Principal Component Analysis* (*PCA*), *Discriminant Analysis* (*DA*) - have proven suitable for vegetable oils authentication.

In this chapter, we aim to investigate the potential of physical methods (NMR or vibrational spectroscopies and chromatography) coupled with statistical data processing to assess vegetable oils authentication issues such as establishing conformity according to label and detection of adulteration.

Keywords: vegetable oils, authentication, adulteration, chemometrics, NMR, FT-IR, multivariate data analysis

1. INTRODUCTION

Vegetable oils are important components of the foods, from multiple viewpoints such as nutritional value, organoleptic characteristics, functional properties within the food matrix. Therefore, the performance of vegetable oils in food applications is dependent of their quality, purity and some intrinsec quality parameters [1], which are in close relationship with oils authenticity. According to a recent study based on both media and scholarly reports [2], vegetable oils and fats (such as olive oils, cocoa butter) are one of the most frequently counterfeited food ingredients or food products, accounting for 24% of the total food fraud cases during 1980-2010 (database available at www. foodfraud.org).

The need for rapid, low-cost and highly accurate analytical methods to assess authenticity and/or to detect adulteration has motivated the use of modern techniques such as NMR or vibrational spectroscopies associated with statistical methods for data analysis (chemometrics) for the characterization of vegetable oils.

Generally, the workflow aiming to establish authenticity or detection of fraud comprises several steps: acquiring an exhaustive collection of authentic oil samples (with known botanical origin, geographical area of production, harvest year, processing techniques), the screening of all samples with specific analytical methods (e.g., spectroscopic, chromatographic, senzorial analyses) in order to obtain a large database regarding oils characteristics. With the aid of statistical analysis (chemometrics), approximately 75% of these data are further used to build up reliable prediction models for oils authentication and/or detection of fraud (adulteration, unconformity with the declared specifications); the remanining 25% of the information in the database is used to test and validate the proposed model, assessing its accuracy. Afterwords, unknown samples may be tested for conformity using the validated model.

Prior to chemometrical analysis, the raw data acquired through analytical methods are usually processed by filtering, feature detection, alignment and normalization [3]. Filtering methods process the raw measurement signals aiming to remove unnecessary and unwanted effects (e.g., baseline noise drift) or spectral regions (e.g., those corresponding to water or CO_2 absorptions in FT-IR spectra). Feature detection is used to detect representations of measured ions from the raw signals. Alignment methods cluster measurements across different samples, while normalization removes embarassing systematic variation among samples.

Having prepared the acquired raw data, it is afterwords necessary to analyse the huge amount of information (variables and features) obtained from the database of the investigated authentic samples. This results in multivariate data matrices that require the use of advanced mathematical and statistical procedures aiming to efficiently extract the maximum useful information from the data. There are two types of multivariate data analysis methods: unsupervised and supervised. The most common unsupervised method is *principal components analysis* (PCA) [3].

2. General Aspects Regarding Physical Methods of Oils Analysis

2.1. General Aspects and Interpretation of Vibrational Spectra (FT-IR, NIR, Raman) of Oils

a. FT-IR Spectra

The FT-IT spectroscopy is a very important technique for both molecular structure elucidation of oils and authentication purposes, its popularity being due on one hand to the fact that FT-IR spectra of vegetable oils provide useful information with easily assignable bands to specific functional groups, and on the other hand to other advantages such as rapidity, no sample preparation required prior to analysis and relatively low price of equipment.

FT-IR spectra of vegetable oils have the same shape (Figure 2.1), each band or "shoulder" giving structural information on functional groups in lipids or impurities present in oils [4].

If the sample contains traces of water, hydroperoxides (R-OOH) or alcohols (from the degradation of hydroperoxides), the stretching vibrations of -OH bonds will appear as broad bands in the 3700-3400 cm^{-1} region.

The 3025-2850 cm^{-1} spectral region is dominated by the intense absorption bands due to the C-H stretching vibrations from methylene and terminal methyl groups of the aliphatic fatty acyl chains in triglycerides [4]; the stretching vibrations of C-H from *cis* CH = CH double bonds are observable at the upper end of this region, while the absorptions of *trans* double bonds appear at higher frequencies, as a very weak band.

At the lower end of this region, carbonyl compounds (resulting from oils oxidation) give a weak absorption. The very intense band at 1740 cm^{-1} is due to the stretching vibration of carbonyls from triglyceride esteric bonds. Free fatty acids (R-COOH) occurring from lipolysis will generate a "shoulder" on the low-frequency side of this band. Carbonylic compounds (aldehydes and ketones) occurring from autooxidation of vegetable oils give specific absorptions in the 1730-1650 cm^{-1} region.

The next region (ranging from 1500 to 900 cm^{-1}) is called the "fingerprint" region because it has a distinctive pattern for any pure compound. Isolated *trans* C = C bonds give a specific absorption band at 966 cm^{-1} due to the C-H out-of-plane deformation; conjugated *trans-trans* or *cis-trans* systems give absorptions at slightly higher frequencies. The C-H out-of-plane deformation of *cis* C = C appears as a "shoulder" at the low-frequency side of the band at 723 cm^{-1} (assigned to the CH_2 rocking vibration) [4].

Overall, mid-FTIR spectroscopy is useful as analytical technique for the rapid and accurate characterization of food samples, especially in identifying specific functional groups present in vegetable oils based on the fact that different functional groups absorb different frequencies of radiation. A part of the infrared spectrum ranging from 650 to 1500 cm^{-1} is known as the "fingerprint" region and is unique to each molecule, allowing for identification of similar substances.

b. FT-Raman Spectra

Raman spectroscopy is based on inelastic scattering of a monochromatic light, which interacts with molecular vibrations or other excitations in the sample, resulting in shifted energy frequencies. The shift in energy frequencies gives information about the vibrational modes in the sample, and consequently spectral bands can be assigned to specific functional groups [5].

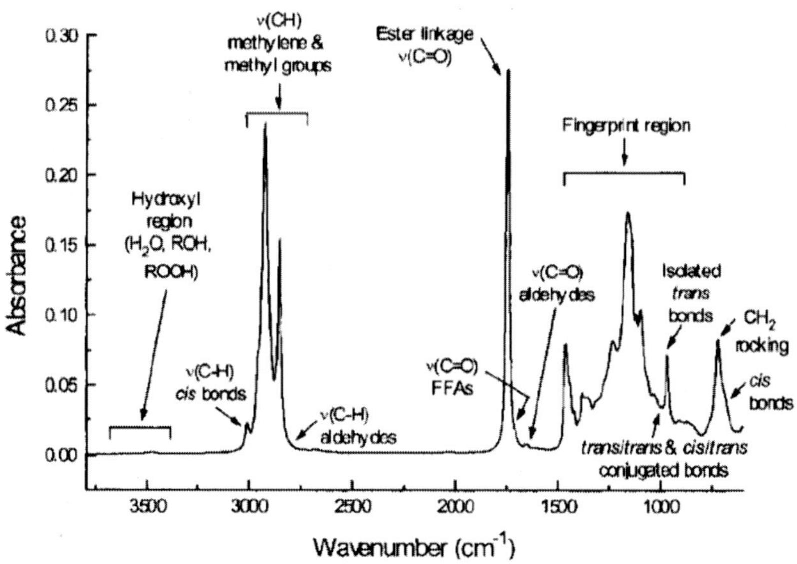

Reprinted from Ref. [4]. Copyright (2001) with permission from WILEY-VCH.

Figure 2.1. FT-IR band assignments for a typical edible oil spectrum.

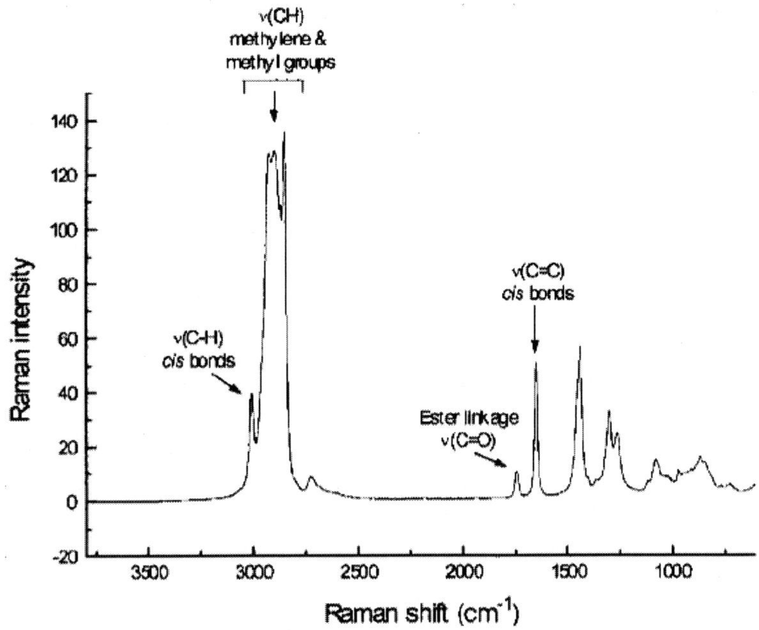

Reprinted from Ref. [4]. Copyright (2001) with permission from WILEY-VCH.

Figure 2.2. Typical FT-Raman spectrum of an edible oil.

Figure 2.2 shows a typical FT-Raman spectrum of a vegetable oil [4]. Compared to *Figure 2.1*, the relative intensities of IR and Raman bands differ considerably, the information provided being complementary.

The 1700-1600 cm^{-1} spectral region has a particular importance for the characterization of oils from the unsaturation point of view. The dominant features in this region are the intense absorption bands corresponding to C=C bonds, in contrast with FT-IR spectra, where the equivalent absorptions are displayed as weak bands [4].

Moreover, in Raman spectra the C = C stretching vibrations corresponding to the *cis* and *trans* configurations appear at different wave numbers (1656 and 1670 cm^{-1}, respectively), therefore Raman spectroscopy is considered a valuable analytical method in the oils field for the determination of the unsaturation type and degree.

c. NIR Spectroscopy

Near infrared (NIR) spectroscopy uses the near-infrared region of the electromagnetic spectrum (frequencies ranging from 800 to 2500 nm) and is based on molecular overtone and combination vibrations [6].

In contrast with FT-IR and Raman spectra, NIR spectra of oils give few structural information, therefore the NIR bands cannot be specifically assigned to functional groups. As it is reflected from *Figure 2.3*, the NIR spectrum of a vegetable oil – recorded between 10000 and 4500 cm^{-1} – displays three broad bands resulted from overlapping overtones and combinations of the fundamental vibration modes associated with spectral IR and Raman bands.

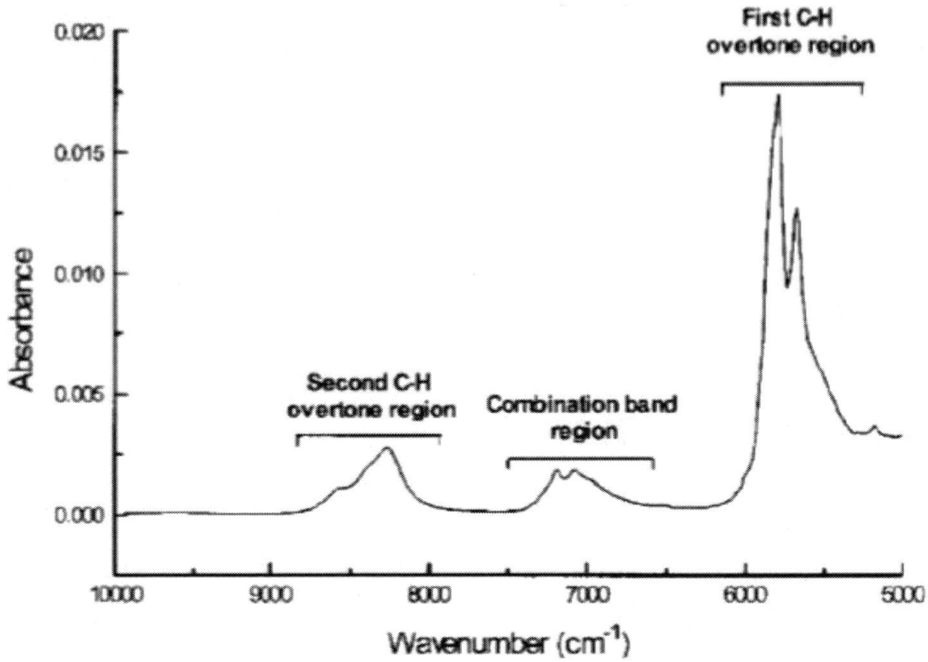

Reprinted from Ref. [4]. Copyright (2001) with permission from WILEY-VCH.

Figure 2.3. Typical NIR spectrum of an edible oil.

Despite the apparent lack of information, NIR spectroscopy is useful as a fingerprint screening technique for authentication purposes of oils and fats. As a relevant example, a FT-NIR spectral data-based method was adopted by the American Oil Chemists' Society (AOCS) as an official method for the determination of the oils unsaturation index (iodine number) [7].

2.2. General Aspects and Peak Assignment of ^{1}H- and ^{13}C-NMR Spectra of Oils

^{1}H-NMR spectra of vegetable oils have the same shape, they differ only by the integrals' values and signal intensities. *Figure 2.4* presents for exemplification the typical ^{1}H-NMR spectrum of a vegetable oil (soybean) and *Table 2.1* lists the chemical shifts and the peak assignment [8].

Compared to ^{1}H-NMR spectra, the electronic configuration of the ^{13}C nucleus determines the spreading of the chemical shifts over ~200 ppm frequency range, resulting in minor signal overlapping and, as a consequence, enabling more refined structural information allowing to determine the positional distribution of the acyl chains on the glycerol backbone [9, 10].

Besides expensive analytical equipment, the major drawback of the ^{13}C-NMR spectroscopy is due to the low natural abundance of ^{13}C and a small gyromagnetic ratio, which explain the low sensitivity of the method (only major components such as triglycerides can be quantitatively analyzed). However, quantitative information from ^{13}C-NMR spectra can be obtained with proper modification of the pulse sequences and acquisition parameters [10].

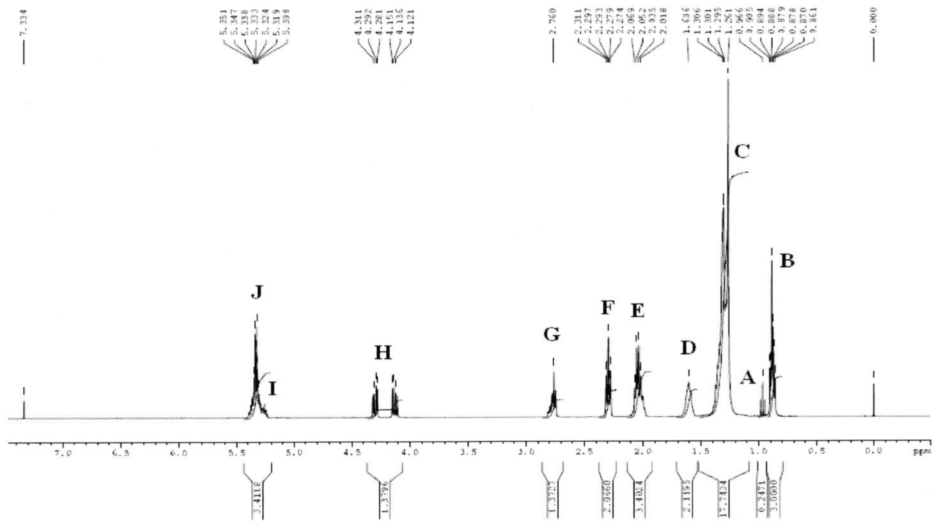

Figure 2.4. Typical 1H-NMR spectrum of a vegetable oil (soybean).

Table 2.1. Chemical shifts and peak assignment of ^{1}H-RMN spectra of vegetable oils [8]

Signal	δ (ppm)	Proton	Compound
A	0.95	-CH=CH-CH$_2$-*CH*$_3$	linolenic acid
B	0.85	-CH$_2$-CH$_2$-CH$_2$-*CH*$_3$	all acids except linolenic acid
C	1.2	-(*CH*$_2$)$_n$-	all fatty acids
D	1.6	-*CH*$_2$-CH$_2$-COOH	all fatty acids
E	2.02	-*CH*$_2$-CH=CH-	all unsaturated fatty acids
F	2.2	-*CH*$_2$-COO-	all fatty acids
G	2.76	-CH=CH-*CH*$_2$-CH=CH-	linoleic acid and linolenic acid
H	4.19	-*CH*$_2$OCOR	all fatty acids
I	5.15	-*CH*OCOR	all fatty acids
J	5.29	-*CH*=*CH*-	all unsaturated fatty acids

Fatty acid methyl esters (FAME) obtained by transesterification of vegetable oils have characteristic ^{13}C-NMR spectra, differing from one to another by integrals' values and signal intensities, as shown in *Figure 2.5.*

Acquisition parameters of the ^{13}C-NMR spectra involves a pulse angle of 30º and a relaxation delay of 30 s in order to confer spectra a semi-quantitative character. A number of 64 scans may be used for a reasonable experimental time. A greater number of scans, although leading to an improvement of the integration quality, is not suitable because it considerably increases the time of acquisition [11].

It can be noticed from the spectra that carbon signals group into five main spectral regions: a) carbonyl atoms region (~175 ppm); b) carbon atoms involved in -C=C- double bonds region (~130 ppm); c) carbon atoms of the methyl ester groups region (~50 ppm); d) methylene groups region (~20-35 ppm); e) chain ending methyl groups region (~14 ppm).

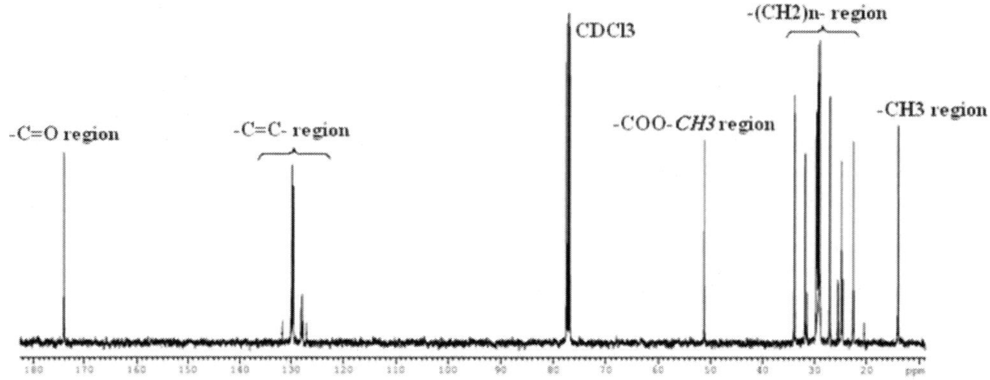

Figure 2.5. ^{13}C-NMR spectrum of a typical vegetable oil FAME.

A particularly interesting region for the quantification of component FAME is the methylene groups region (*Figure 2.6*). Despite the large number of signals in this region, they could be resolved on the basis of literature information [9, 10, 12] and chemical shifts computation based on increment additivity [13].

Another spectral window with practical utility for the chemometric determination of oils composition is the region specific for the unsaturated carbon atoms involved in –C=C– bonds in the fatty acyl chains (~130 ppm), presented in *Figure 2.7*. Peak assignments for this region could be resolved (*Table 2.2*).

In the cases where signal overlapping occurred, the entire group of signals may be integrated, in order to comply with the integration rule, from baseline to baseline (N, M or G groups). In such conditions, a system of chemometric equations allowing the FAME composition computation in terms of four classes of constituents (linolenic acid, linoleic acid, mono-unsaturated fatty acids and saturated fatty acids) on the basis of signals integrals has been set up [11].

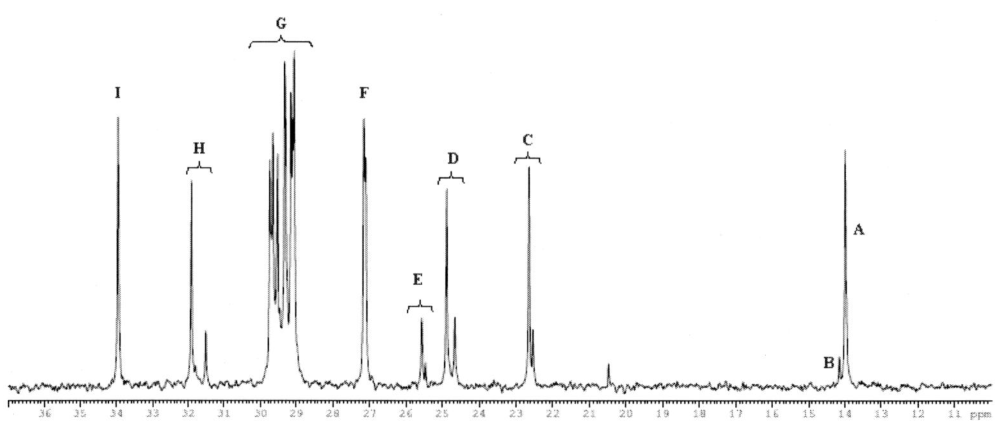

Figure 2.6. 10-37 ppm spectral window of the ^{13}C-NMR spectrum of FAME (corresponding to -$(CH_2)_n$- and –CH_3 groups).

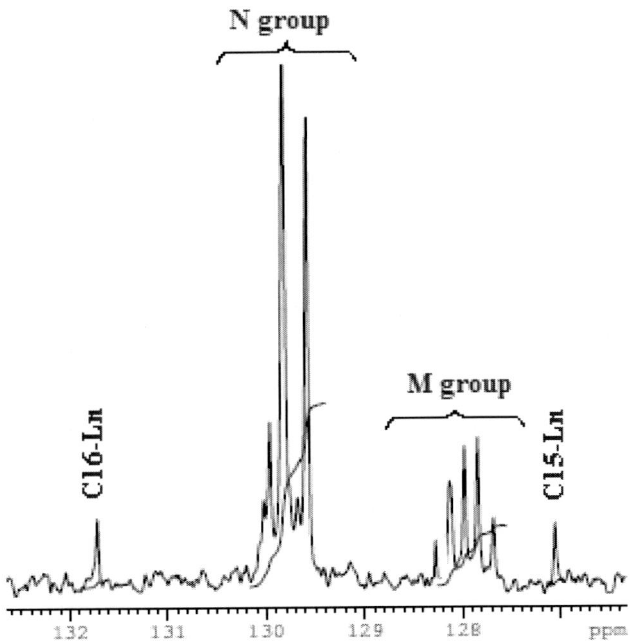

Figure 2.7. 126-133 ppm ^{13}C-RMN spectral window (corresponding to -C=C- region).

2.3. Determination of the Fatty Acids Profile of Vegetable Oils

a. From the Chromatographic Data

Gas chromatography (GC) is the most popular method for the compositional analysis of oils and fats, allowing to obtain accurate and detailed fatty acids profiles.

Prior to GC analysis, lipids must be transformed into the corresponding fatty acids methyl esters (FAME), more volatile than triglycerides or other complex lipids. The choice of the most convenient lipid derivatization method is crucial for the relevance of the composition results. FAME may be obtained through transesterification of triglycerides with methanol either under alkaline or acidic catalysis [14]. In both cases, the transformation involves two steps: saponification (usually carried out with NaOH), followed by the reaction with MeOH, in the presence of the catalyst. In the case the alkaline catalysts (sodium methoxide being preferred) are used for methylation, the free fatty acids and the fatty acids from sphingolipids remain unesterified. The sodium methoxide-based esterification method, even if it is not effective for the free fatty acids, is however very rapid: it was demonstrated that triglyceride derivatization into the corresponding FAME is completely achieved within 2-5 minutes [15].

The most intensively used acid catalysts for lipid esterification purposes are the BF_3-methanol complex (10-14% m/vol.) [14], BCl_3-methanol (5% m/vol.), anhydrous HCl in methanol [16] and H_2SO_4 in methanol (1-2%). The advantage of acid catalysis over the alkaline conditions is that it allows a rapid esterification of the free fatty acids; therefore, if saponification is performed prior to methanolysis under acidic conditions, the method becomes effective for the complete lipid derivatization into the corresponding FAME.

Table 2.2. ^{13}C-RMN peak assignments for FAME (observed and calculated* values)

No.	Signal	(δ, ppm)	Peak assignment	Calculated* chemical shift				
				Methyl palmitate	Methyl stearate	Methyl oleate	Methyl linoleate	Methyl linolenate
1.	K	174.1	C-1	173.1	173.1	173.1	173.1	173.1
2.	-	131.7	C-16 (Ln)	-	-	-	-	128.9
3.	Group N	130.1	C-9 (Ln)	-	-	-	-	132.2
4.		130.0	C-13 (L)	-	-	-	132.2	-
5.		129.9	C-10 (O)	-	-	130.7	-	-
6.		129.8	C-9 (L)	-	-	-	132.2	-
7.		129.6	C-9 (O)	-	-	130.7	-	-
8.	Group M	128.3	C-12 (Ln)	-	-	-	-	28.8
9.		128.1	C-13 (Ln)	-	-	-	-	128.8
10.		127.9	C-10 (L)	-	-	-	127.3	-
11.		127.8	C-12 (L)	-	-	-	127.3	-
12.		127.7	C-10 (Ln)	-	-	-	-	127.3
13.	-	127.0	C-15 (Ln)	-	-	-	-	128.0
14.	J	51.2	-COO-*CH*$_3$	51.9	51.9	51.9	51.9	51.9
15.	I	33.9	-O-CO-*CH*$_2$-	33.6	33.6	33.6	33.6	33.6
16.	H	31.5-31.9	-*CH*$_2$-CH$_2$-CH$_3$	31.9	31.9	31.9	32.0	-
17.	G	29.0-29.7	-(CH$_2$)$_n$-	29.7	29.7	29.7	29.7	29.7
17.	F	27.1	-CH=CH-*CH*$_2$-	-	-	33.7	33.8	33.8
19.	E	25.4-25.6	-CH=CH-*CH*$_2$-CH=CH-	-	-	-	37.6	37.7
20.	D	24.6-24.9	-CO-CH$_2$-*CH*$_2$-	25.1	25.1	25.1	25.1	25.1
21.	C	22.5-22.7	-*CH*$_2$-CH$_3$	22.8	22.8	22.8	22.9	-
22.		20.5	-HC=CH-*CH*$_2$-CH$_3$ (Ln)	-	-	-	-	26.6
23.	B	14.2	-CH$_2$-*CH*$_3$ (Ln)	-	-	-	-	14.3
24.	A	14.0	-CH$_2$-*CH*$_3$	14.1	14.1	14.1	14.1	-

* Based on additive (incremental) computations [8, 13]; Ln – linolenic acid; L – linoleic acid; O – oleic acid; S – stearic acid; P – palmitic acid.

The configuration of the GC system also affects the result of the chromatographic analysis. Many factors, such as the column type, gas flow control, and temperature programming – if not set up correctly – will affect the performance of the GC. The chromatographic columns represent the critical parameter of a GC system, their choice depending on the nature of the analyzed sample. There are many GC columns available for FAME analysis. Generally, a good column for FAME should be able to separate all the components and allow the FAME to elute primarily according to carbon chain length and secondarily by the number of double bonds. There should be no overlapping or minimal overlapping among FAME having different chain lengths [14]. Generally for FAME analysis, polar and high-polar columns are preferred because of the presence of unsaturated fatty acids and their positional isomers which have high polarity as compared to their saturated counterparts.

A special issue is the GC separation of the *cis-trans* or positional isomers of unsaturated fatty acids methyl esters. It can be resolved if very long capillary columns are used for the separation. Performant GC capillary columns allow both specific identification and quantitative determination of the *trans* FAME [14].

Once introduced into the column, FAME with different carbon chain length and saturation levels move through the column at different rates and elute from the end of the column sequentially. Fatty acid standards (Nu Chek Prep, Matreya, *Supelco* 37 FAME Mix) are required to obtain the retention time for individual fatty acids in GC analysis, so that fatty acids in samples can be identified by comparing their retention times with those of the standards. An example of the FAME identification by comparing the retention time of each chromatographic peak with the peaks of a commercial standard FAME mixture is shown in *Figure 2.8.*

b. From the ¹H-NMR Data

Based on ¹H-NMR spectral data, a system of chemometric equations leading to the determination of oils composition on four classes of fatty acids (linolenic acid, linoleic acid, mono-unsaturated fatty acids and saturated fatty acids) was developed [17], the obtained results being in agreement with the composition determined by GC:

$$x = \frac{I_A}{I_A + I_B}\,;\; y = \frac{I_G - 4kx}{2k}\,;\; z = \frac{I_E}{4k} - x - y\,;\; t = 1 - x - y - z$$

where:

- x, y, z and t represent the molar ratios of linolenic acid, linoleic acid, mono-unsaturated fatty acids (oleic acid), and saturated fatty acids, respectively;
- I_A, I_B, I_E, I_F, I_G represent the integral values of the corresponding signals.
- k is a spectrometer constant; it is a computed coefficient which correlates the signal integral with the number of protons generating the signal. By consequence, k value can be determined on the basis of different signals, the most accurate results being obtained for. $k = \frac{I_F}{2}$ [17].

The fatty acids composition of vegetable oils has also been estimated by ¹H-NMR spectroscopy on the basis of the relationship between the integrals and the characteristic signal for each fatty acid and the signals of the glycerol moiety [18] with four integrations and three integral subtractions. The results were found in agreement with those obtained by the conventional method (GC).

It has been previously shown (Section 2.3.a) that gas-chromatographic methods allow structural and compositional analysis of vegetable oils. These methods require sample derivatization (triglycerides) into the corresponding more volatile fatty acids methyl esters (FAME) prior to analysis. Besides being a tedious step, there is always a risk of modification of the original fatty acids composition during the separation or reflux involving steps (isomerization) in the analysis protocol. In terms of time length, not only the sample preparation prior to analysis is time consuming, but even the GC separation itself, a typical chromatogram being recorded in 45-50 minutes. Overall, a gas-chromatographic analysis of a vegetable oil sample takes about two hours. But, of course, precision – the main advantage of the GC method – should never be neglected.

On the other hand, NMR spectroscopy (¹H- or ¹³C-) allows for simultaneous structural and compositional analysis of complex mixtures such as vegetable oils on four classes of fatty acids (saturated, mono-, di and tri-unsaturated) with certain advantages, such as short acquisition times (a typical ¹H-NMR spectrum takes about two minutes to be recorded), no sample preparation prior to analysis (it is applicable directly on triglycerides), the method being non-invasive and allowing complete sample recover after analysis. The major drawback of the NMR techniques compared to gas-chromatography is the prohibitive price of the instruments.

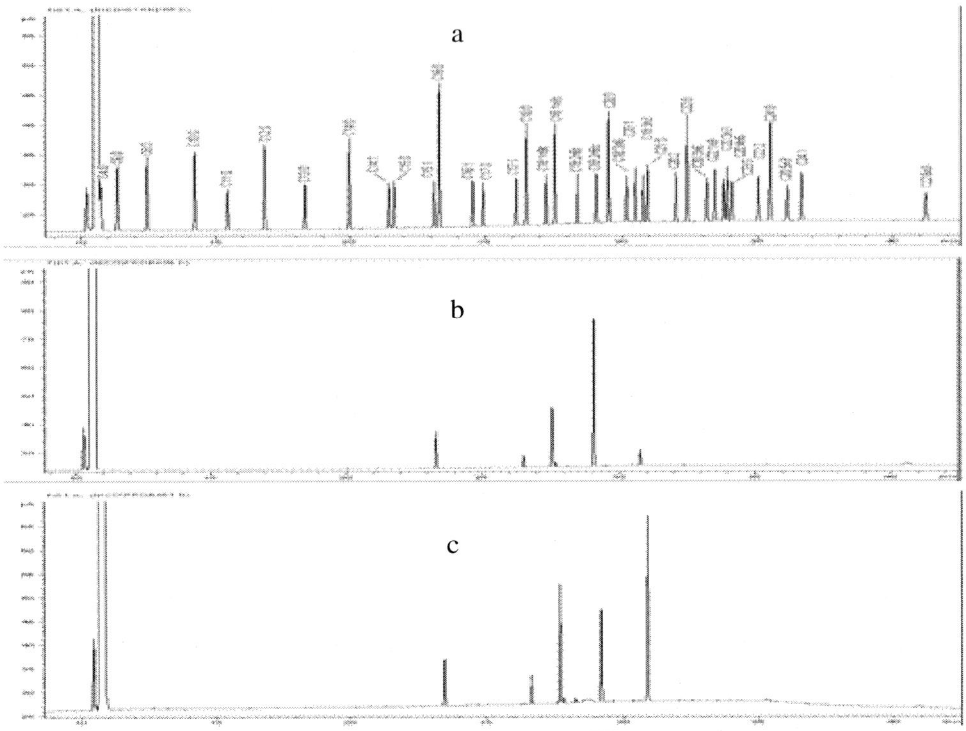

Figure 2.8. FAME identification: a) standard FAME mixture chromatogram (*Supelco* 37 FAME Mix); b) FAME from soybean oil; c) FAME from linseed oil.

2.4. Statistical Data Analysis

The modern physical and chromatographic methods for food analysis in general (and oils analysis in particular) produce huge amounts of data; data analysis not only deals with the extraction of primary information from data, but also with the generation of secondary information, for example the generation of models which can further be used for prediction purposes. Such huge amounts of data can be investigated through statistical methods. For oils authentication purposes, the main statistical tools are Principal Components Analysis, Partial Least Squares and the analysis of variance, which will be briefly described below.

1. *Principal Component Analysis (PCA)* is an unsupervised modelling method allowing exploratory data analysis and the classification of samples by investigating similarities and differences between different samples [19]. The PCA technique reduces the number of latent variables, called *principal components* (PC). Each PC explains a part of the system's variability of the original data set. PCs are ranked as PC1, PC2, PC3 etc. in descending order in terms of the amount of information. One of the significant goals of PCA is to eliminate the principal components associated

with noise, thereby reducing the dimensionality of complex problems and minimizing the effect of measurement errors [20]. By reducing the data dimensionality, PCA allows its visualization while retaining as much as possible the information present in the original data. In other words, PCA transforms the original measured variables into new uncorrelated variables, called principal components.

It is a common practice to orthogonally represent two PCs relative to each other in order to obtain plots (generally PC1/PC2 plots) in which samples group according to their similarities or differences, allowing the authentication of oil type, oil variety, geographical area of production, harvest year and even the detection of adulteration. PCA is suitable as statistical method for spectroscopic and chromatographic data analysis.

For example, oils of different botanical origin (conventional and high-oleic varieties of sunflower, rapeseed, soybean and linseed) were classified on the basis of the principal component analysis (PC1/PC2 score plot representation) of compositional data on four components (linolenic acid, linoleic acid, mono-unsaturated and saturated fatty acids, respectively) computed from ^{1}H-NMR data [21]. The PC3/PC2 score plot representation allowed for the authentication of the geographical area of production of the rapeseed oil samples, while samples of soybean oil grouped according to the harvest year (2005, 2006 or 2007) in the PC1/PC2 score plot.

More recently, researchers used three dimensional representation of the PC scores (PC1/PC2/PC3) to authenticate the harvest year for olive oil samples from Greece with a 94% correct classification, reflecting the significant year-to-year variability of the fatty acids profiles.

2. *Multiple Linear Regression (MLR)* models a linear relationship between a dependent variable and one or more independent variables.
3. *Principal Component Regression (PCR)* is a combination of PCA and MLR: the scores gained by principal component analysis are used for multiple linear regression.
4. *Partial least squares (PLS)* regression is a supervised method based on the relationship between the signal intensity in a spectrum and the characteristics of the sample. The method is useful in the case of overlapping signals.

   The algorithm is based on the mathematical correlation of spectral data to a property matrix of interest while simultaneously accounting for all other significant spectral factors that perturb the spectrum. Therefore, PLS is a multivariate regression method that uses the full selected spectral region. PLS is a linear regression extension of PCA, which is used to connect the information in two blocks of variables X and Y to each other. It can be applied even if the features are highly correlated [22].

   PLS can be applied to establish a predictive model, even if the features are highly correlated. The method aims to establish a relationship between two matrixes X and Y (X = data matrix and Y = properties matrix). The procedure involves the following steps: the principal components are calculated for X and Y matrixes, separately; the scores of the X matrix are then used for a regression model to predict the scores of Y, which can further be used to predict Y. A crucial decision in PLS is the choice of the number of principal components used for the regression. A good approach to solve this problem is the cross-validation [22].

5. *Linear Discriminant Analysis (LDA)* is a method of finding such a linear combination of variables which best separates two or more classes. In itself LDA is not a classification algorithm, although it makes use of class labels. However, the LDA result is mostly used as part of a linear classifier. The other alternative use is making a dimension reduction before using nonlinear classification algorithms. It should be noted that several similar techniques (differing in requirements to the sample) go together under the general name of Linear Discriminant Analysis.
An example of this procedure is illustrated in *Figure 2.9* [23] showing a very good separation into clusters of oil samples from virgin olive oil, hazelnut, sunflower, soybean and corn oils, respectively on the basis of Linear Discriminant Analysis applied to ^{1}H and ^{31}P NMR metabolic profile data. After appropriate validation, the model was used to detect adulteration using admixtures of virgin olive oils with the above mentioned oils.

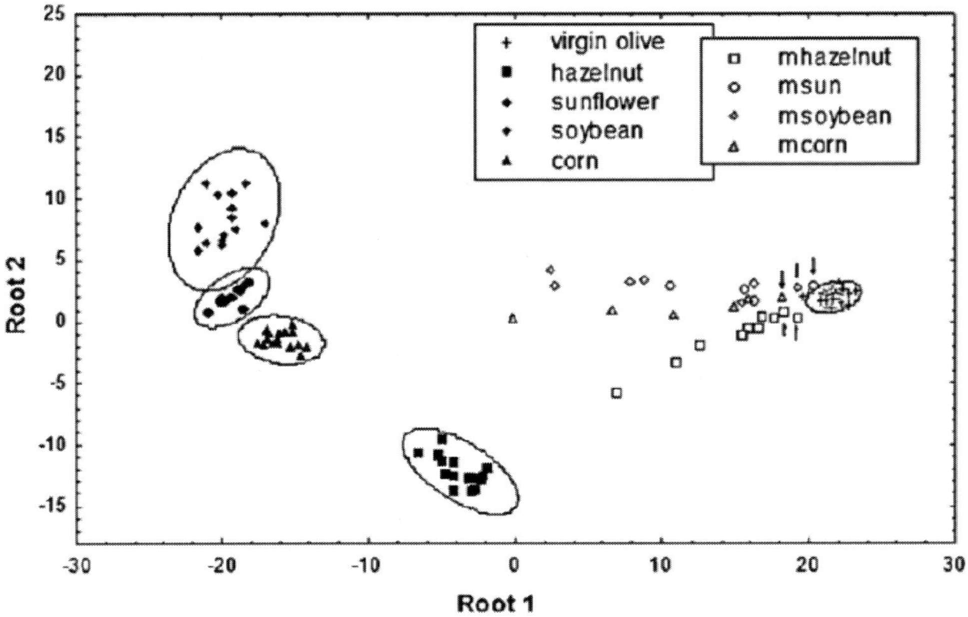

Figure 2.9. Plot of discriminant functions roots 1 and 2 for five types of edible oils. Virgin olive oil, soybean, hazelnut oil, corn oil and sunflower oil are shown by crosses and solid symbols. Four sets of mixtures of 5, 10, 15, 20, 35 and 50% (w/w) of virgin olive oils with hazelnuts (mhazelnut), sunflower (msun), soybean (msoybean), and corn (mcorn) oils and one set of mixtures of 5, 10, 15 and 20% of virgin olive oils with hazelnut oils (mhazelnut) are denoted by open symbols. Arrows indicate mixtures of 5% (w/w) of seed oils in virgin olive oils.

6. *Analysis of variance (ANOVA)* tests if one group of samples differs from the population of samples investigated [24]. Multiple measurements are necessary to establish a benchmarked variability ("within-group") characteristic for the sample type. Whenever a difference significantly exceeds this benchmark, at least two populations of samples are involved.

7. *Cluster Analysis (CA)* aims to divide a group of objects into clusters so that the objects within a cluster are similar, but objects taken from different clusters are dissimilar. Hierarchical clustering methods organize samples into clusters of increasing size, with small clusters of related samples being grouped together into larger clusters: at one extreme each sample is in a separate cluster; at the opposite extreme, all the samples are in one single cluster. The relationships between the clusters can be visualized using a dendrogram. Cluster analysis is sometimes referred to as a distance-based approach to sample selection since it requires a measure of the "distance" between pairs of samples. This distance is often calculated as $1\text{-}S$ (S being the similarity coefficient) when samples are represented by physico-chemical properties (or their scaled derivatives or a set of principal components) [25].

3. Authentication of Vegetable Oils

There is a multitude of factors influencing the chemical composition of oils; if their importance is underestimated, they may obscure the interpretation of the authentication results. One of the critical steps in calibrating a reliable classification/prediction model for authentication purposes is to build up an exhaustive collection of authentic oil samples with known botanical origin (plant species), variety, geographical cultivation area and year of harvest. Ideally, the samples must represent all the factors – or at least the most important ones – that have influence on the composition of oils. The major factors affecting the composition of vegetable oils are the pedoclimatic conditions (such as soil, altitude, climate), genetic factors (variety), agronomic practices (*e.g.*, irrigation), ripeness which is closely correlated with the month of harvest (e.g., from October to February in the case of olives) or storage time and storage conditions before and after oil extraction [26]. For the different vegetable oils composition, the influence of these factors was examined in several instances with both conventional and modern techniques. A variety of metabolites and/or spectral regions of NMR or vibrational spectra (FT-IR, NIR, Raman) coupled with a variety of statistical methods for data analysis (chemometry) have been applied for the classification of the collected oil samples according to plant species, variety, geographical area of production or year of harvest, with different classification rates. Concerning the olive oil, these issues are presented in some recent excellent review articles [26-28].

3.1. Authentication of the Plant Species

Infrared (IR) and Raman spectroscopies were used as sample screening techniques for the authentication of vegetable oils according to their botanical origin (plant species) [29]. An exhaustive data base consisting of authentic samples of olive (extra-virgin and refined), sesame, rapeseed, sunflower, soybean, mustard, peanut, palm, walnut, corn, coconut, palm kernel, sweet almond, grape seed, safflower and hazelnut oils was used to obtain spectral data which were further chemometrically classified and differentiated on the basis of the linear discriminant analysis. IR spectral data gave 100% classification rates, being more effective compared to Raman or IR-Raman combined data.

Fourier transform (FT) vibrational spectroscopic techniques (NIR, mid-IR and Raman) were used for the discrimination among ten different edible oils and fats, both vegetable (extra virgin olive oil, canola, safflower, corn, soybean, peanut, and coconut oils) and animal (lard, butter, cod liver oil) [30]. The acquired spectral data were statistically analyzed with specific methods (linear discriminant and canonical variate analyses) for classification and discrimination purposes. The most efficient screening technique was FT-IR spectroscopy, which gave 98% accuracy in the classification of oils and fats according to their biological origin. FT-Raman and FT-NIR proved also efficient, with classification accuracies of 94 and 93%, respectively. More recently (in 2013), FT-IR spectroscopy was once again found efficient for the discrimination of different plant oils (soybean, canola, sunflower, corn and olive) [31], on the basis of canonical variate analysis (CVA) and partial least squares-discriminant analysis (PLS-DA). CVA classification model showed 100% sensitivity and specificity for both calibration and external validation purposes.

^{1}H-NMR in combination with PCA allowed the differentiation of Romanian seed oils from four plants (sunflower, soybean, rapeseed and linseed) [21]. The score plot representation of PC1/PC2 loadings of the samples from the calibration group revealed a complete separation of oils into four classes, corresponding to their botanical origin. The proposed model for oils differentiation according to the plant species was tested with samples from the test group; all these samples were properly assigned to the corresponding botanical origin, thus demonstrating the efficiency of the method. An even easier strategy for the vegetable oils authentication (sunflower, soybean, rapeseed and linseed) based on the ^{1}H-NMR data uses the integral values of seven selected peaks in the oil spectrum [32] processed by principal components analysis.

^{1}H-NMR and FT-IR spectral data were used for the compositional characterization of the grape seed oils (extracted from authentic grape seeds or commercial samples) and in authenticity control studies [33]. The NMR analysis was conducted directly on the grape seed oils (as triglycerides), without any other sample derivatization. The compositional characterization was based on a system of chemometrical equations [17] allowing the determination of the fatty acids profiles in terms of tri-unsaturated fatty acids (linolenic acid), di-unsaturated fatty acids (linoleic acid), mono-unsaturated fatty acids (oleic acid) and saturated fatty acids (palmitic and stearic acids) using the ^{1}H-NMR data. The compositional differences between oils from grape seeds and common genuine oils (sunflower, soybean, linseed and rapeseed) were investigated and the potential application in authenticity control was assessed based on the ^{1}H-NMR and FT-IR spectral data analyzed with statistical methods (PCA). A good separation into specific clusters was obtained for the authentic vegetable oils on the basis of the ^{1}H-NMR data (*Figure 3.1*). Moreover, a good separation of Burgundy, Cabernet Sauvignon and Muscat Ottonel varieties from the rest of the grape seed oils was obtained. Additionally, a tendency of grouping according to red grape or white grape varieties was observed, clusters being placed in different quadrants of the score plot (*Figure 3.2*). Commercial grape seed oils (sample C1, C2, C3 and C4), although situated within the grape seed oils area, were still very close to the specific area of sunflower oils.

The compositional differences between oils from grape seeds and common oils using FT-IR spectroscopy were established by statistical analysis, PCA being considered as classification method [33]. Differentiation of the samples was performed using the relative intensity of absorption bands. The peak intensities of the absorption bands corresponding to the main classes of chemical compounds identified in the IR spectrum were measured.

The area between 3050 and 4000 cm^{-1} was eliminated from the study because it contains information which is not relevant for oils discrimination (water absorbance), being also a source of noise in the spectrum. From the 2650–3050 cm^{-1} spectral range, nine values of the absorption bands intensities (every 50 cm^{-1}) were obtained. In the same way another nine values from 1600-2000 cm^{-1} and 19 values between 600-1500 cm^{-1} spectral range were obtained. In total, each IR spectrum was represented as a vector with 37 values. A good separation of the grape seed oils from the other genuine common oils is shown in *Figure 3.2*. Samples cluster in a well-defined group, separated from the specific areas of the rest of the oils.

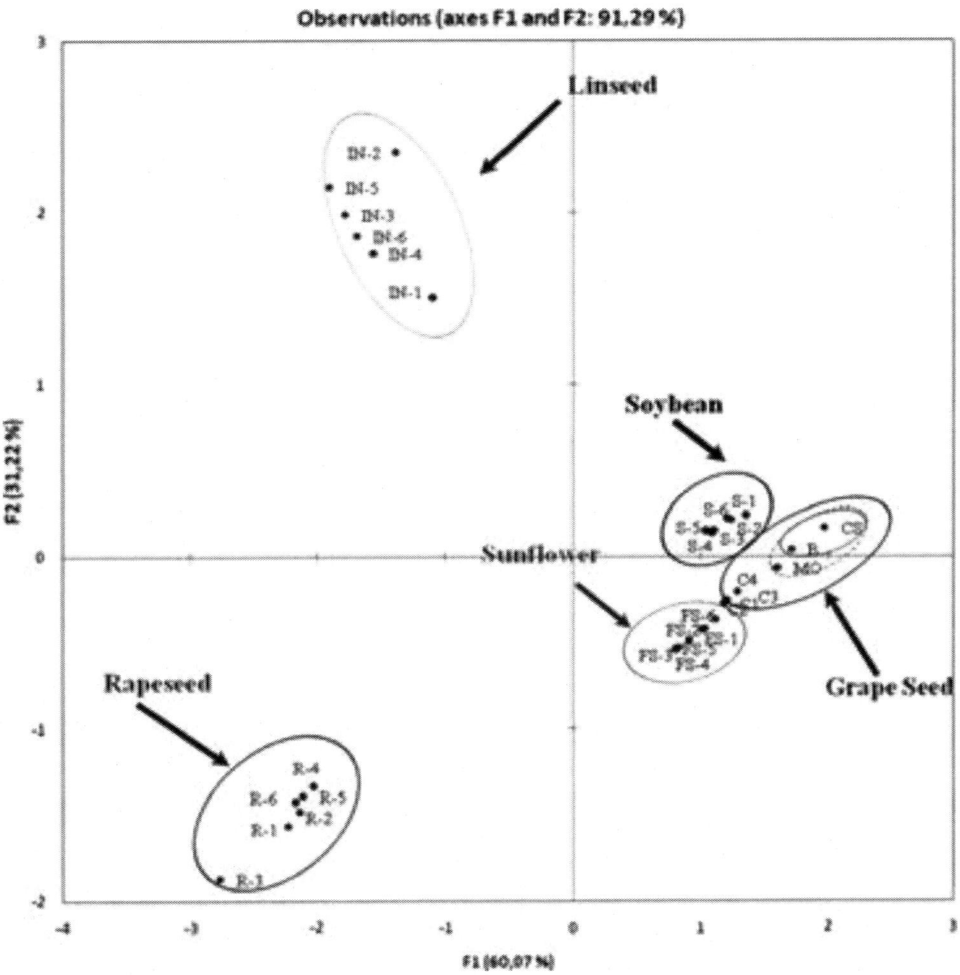

Figure 3.1. Principal component (factor) scores F1/F2 plot for the grape seed oils, sunflower oil, rape soybean oil, and linseed oil. Samples analyzed through ^{1}H-NMR and PCA.

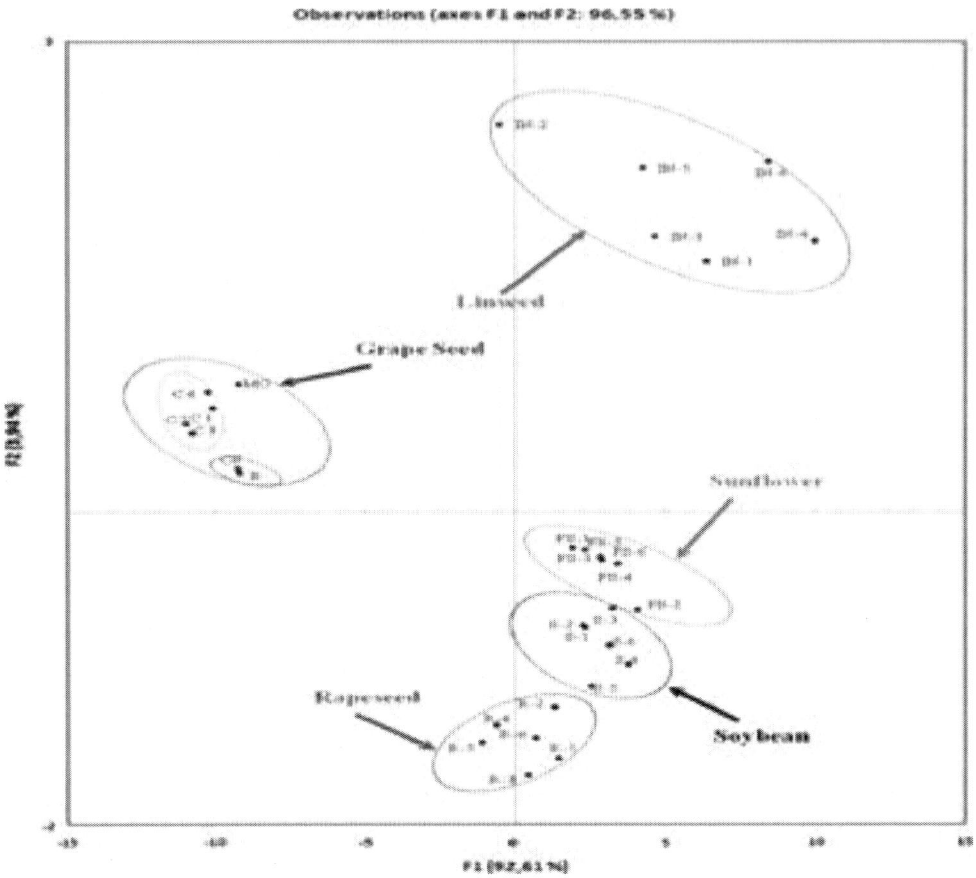

Figure 3.2. Principal component (factor) scores F1/F plot for the grape seed oils, sunflower oil, rapeseed oil, soybean oil, and linseed oil. Samples analyzed through FT-IR and PCA.

3.2. Authentication of the Plant Part

In some cases (olive or palm fruits), the oil can be extracted from different parts of the plant, namely the pulp and the kernel. Generally, the most appreciated are oils obtained from the pulp, having superior nutritional value in comparison with the kernel oils which are considered as by-products.

In the olive oil industry, the de-stoner technique is currently the only mechanical system that allows a selective crushing of the fruits excluding the kernel from the paste [34]. The kernel exclusion has the certain advantages to increase both nutritional quality of the product and sustainability of the process. If the olive paste is made exclusively from the mesocarp, the polyphenols content is protected from oxidation because most of the oxidative enzymes, as polyphenol oxidase and peroxidase (mainly found in the kernel) are removed [35-38]. On the contrary, when the "integral crushing" is applied, the presence of olive kernels reduces the work capacity of the whole mill plant; in addition, the crushing of the kernel process also involves a considerable amount of energy dissipated as thermal energy which heats the olive paste before the triggering of thermolabile enzymes, such as lipoxygenase and hydroperoxide lyase [34, 36]. The commercial market does not allow kernels and extra charge the oil price to compensate for the loss of yield. Consequently, the development of analytical techniques able to ascertain that the oil from olives pitted is intact and not adulterated with oil of different origin (not even from kernels from the same plant) is justified by the need of security on product quality since the extra-virgin olive oils have higher market price.

The chemometric discrimination of extra virgin olive oils obtained from whole and stoned olive pastes from four Italian commercial brands was carried out by using FTIR spectroscopy and partial least squares-discriminant analysis (PLS-DA) approach [34]. The adopted chemometric methodologies were able to describe the different chemical features regarding the phenolic and volatile compounds contained in the two types of oil by using unspecific IR spectral information. PCA was used in cluster analysis to capture data patterns and to highlight differences between technological processes and extra-virgin olive oil brands. Different oil extraction procedures were identified through PLS-DA. Discriminant analysis was extended to assess the possible adulteration by addition of aliquots of oil from whole paste to the most valuable oil from stoned olives with satisfactory external validation of the PLS models.

It was established that the composition of the pulp and kernel oils of macauba (*Acrocomia aculeata*) palm fruit were completely different from each other [39]. The pulp oil contains mainly unsaturated triglycerides (approximately 78.5% of all identified compounds), followed by significant amounts of diglycerides (~ 13.2%) and minor amounts of free fatty acids (~ 5.6%) and sterols (1.5%), while in the case of the kernel oil, saturated triglycerides with shorter chains account for approximately 98.6% of all identified compounds, the remaining being trace amounts of free fatty acids (0.9%), sterols (0.2%) and monoglycerides (0.3%). Saponification of the macauba pulp and kernel oils produced different distribution of fatty acids, with the pulp oil being enriched in unsaturated fatty acids (70%, mainly oleic and linoleic acids), and the kernel oil being enriched in saturated fatty acids (70%, mainly lauric acid). Thermal analyses also indicated different thermal stabilities, with the pulp oil having higher thermal stability than the kernel oil [39].

3.3. Authentication of the Geographical Origin

Authentication of agro-alimentary products is a recurrent concern involving several analytical techniques (chromatographic, spectroscopic, DNA markers, etc.) coupled with statistical methods helping to determine specifications and to identify conformity with known labelled products: Protected Designation of Origin (PDO), Registered Designation of Origin (RDO), Protected Geographical Indication (PGI), etc. These labels represent a protection way of product denomination, since they guarantee that fabrication was performed according to strict and constraining specifications. Therefore, production and processing have to take place in a well-defined geographical area (climate, soil, altitude, etc.) by following a well-recognized and specific know-how. Consequently, these labeled products have more enhanced commercial values compared to similar products [40].

In the vegetable oils field, olive oil is by far the most protected by PDO or PGI legislation.

There are many designations of origin concerning virgin olive oils worldwide, some of them being mono-varietal (*e.g.*, *Nyon*, *Nice*), others being multivarietal (*e.g.*, *Aix-en-Provence*), mixing several compositional varieties (*e.g.*, *Aglandau*, *Grossane*, *Salomenque*). The varieties are required by the specifications concerning the trees grown in the orchards. However, the percentages of olive oil varieties in commercial blends are free under the condition that at least two main varieties co-occur and should be indicated on the product label. Moreover, no minimal percentage of compositional oil variety is fixed such that PDO label-benefiting oil blends are commercialized without such information and their organoleptic characteristics are subjected to significant fluctuations.

Olive oils represent complex matrices varying from pure to heterogeneous varietal contents. Therefore, quantitative analysis of co-occurring components is fundamental for conformity control and adulteration alerting of commercial oils.

No matter the analytical technique used to assess conformity with the declared geographical origin of PDO olive oils, the procedure involves the screening of a large number of samples from different areas and the examination of data with statistical methods to build a classification/prediction model able to correctly assign unknown PDO samples to the corresponding PDO group [26].

The accuracy of the classification models depends largely on the size of the database obtained from the screening of authentic samples. The classification model being based on the oils composition, NMR metabolic profiling and or NMR or FT-IR metabolic fingerprinting appear to be excellent analytical tools to obtain raw data from the authentic samples databases which will further be processed by statistical means.

There are several studies concerning the discrimination of PDO extra virgin olive oils from regional (sites within the same geographical region), national (different area of production within the same country) or even international viewpoint on the basis of the NMR screening. These studies are summarized in an excellent recent review article [26]. As expected, most of these articles focus on different sites of the Mediterranean basin (as the main geographical area for the olive oil production); the PDO international discrimination involves, of course, mainly Mediterranean countries (Italy, Spain, Greece, France, Turkey, Tunisia, Morocco) [41-44].

Stable isotope ratio analysis has proven a useful tool when establishing the geographical origin of vegetable oils. Each biologically important element (C, N, O, H or S) has at least two stable isotopes, the most abundant isotope in a pair being the isotope with the smallest atomic mass: $^{13}C/^{12}C$, $^{15}N/^{14}N$, $^{18}O/^{16}O$, $^{2}H/^{1}H$ and $^{34}S/^{32}S$. It was established that different processes such as photosynthesis, respiration, evaporation, metabolism or organic matter turnover determine variations in isotope abundances.

One of the most studied isotope ratios for vegetable oils authentication is $^{13}C/^{12}C$, mainly because variability in $\delta^{13}C$ values is related to the geographical area of production (altitude, latitude and longitude), year of harvest, climate and the particular variety of oil [45].

PCA technique was used to differentiate hazelnuts (*Tonda Gentile Trilobata* variety) grown in the Piemont (Italy) region from other hazelnuts belonging to different cultivars and/or with a different geographical origin (Chile, Turkey, as well as other Italian regions), based on their chemical composition (fatty acids profiles, polyphenols content, protein patterns), antioxidant activity and genetic characteristics. 3D score plots representation of the first three principal components showed a clear discrimination of hazelnuts on the basis of their geographical origin, all samples being clearly clustered [46].

3.4. Authentication of the Harvest Year

Another factor that may affect the chemical composition of oils is the harvest year. In the case of olive oils, the effect of the harvest year on the metabolites (fatty acids profile, antioxidants or other minor components) has been examined systematically [41, 47]. It was found that the harvest year was the most effective and also the most successful parameter for the discrimination of Greek olive oils, regardless the cultivar or geographical area of production. The significant year-to-year variability of the chemical composition of oils is reflected by the 94% accuracy of the classification. Investigation of year-to-year variability was carried out by external validation of the prediction ability of the classification model obtained from samples harvested in one year with samples harvested in another year [48]. The effect of the harvest year cannot be generalized, since it can be the result of unstable climatic conditions that differentiate the crops characteristics of one year from the other. For example, hot summers or rainy autumns may change the concentration of some oil metabolites. For an accurate classification, a large database from authentic samples obtained from several harvest years is required [26].

NMR spectroscopy and quality parameters such as peroxide value, free acidity, and UV spectrophotometric indices were used to investigate the influence of the harvest period and harvest method on olive oil composition, as well as their interaction with primary (genetic and pedoclimatic) and secondary (agronomic practices and technological procedures) [49]. To avoid confusion, the general linear model analysis was used to adjust ANOVA results and consequently, the effect of the factor of interest was corrected for the effects of the other factors that might influence the variable under investigation. The weight of each factor was then evaluated by the variance component analysis. Chemometry (PCA and LDA) was used to group samples according to the harvest period and harvest method. It was also observed that volatile compounds (hexanal and *trans*-2-hexenal), as well as the *sn*-1,3-diglycerides and squalene significantly decreased during the ripening.

In another study [50], the effect of crop year and harvesting time on the fatty acids composition of virgin olive oil (*Picual* cultivar) was investigated during the fruit ripening period for three crop seasons on the basis of the mean fatty acids composition variation. It was observed that the content of palmitic acid and saturated fatty acids decreased during fruit ripening while oleic and linoleic acids increased, and the amount of stearic and linolenic acids decreased. The saturated acids (palmitic and stearic), as well as the polyunsaturated acids (linoleic and linolenic) were dependent on the time of harvest, whereas the amount of oleic acid varied with the crop year.

The differences observed between crop years for both palmitic and linoleic acids were explained by the differences in the temperature during oil biosynthesis and by the amount of summer rainfall for oleic acid content. A significant correlation was observed between the MUFA/PUFA ratio and the oxidative stability measured by the Rancimat method [50].

3.5. Authentication of the Oil Processing Technique

In the case of olive oils – where virgin olive oils are acknowledged for their incontestable organoleptic and nutritional quality – it is important to establish if oils were obtained only by cold-pressing or if they have been applied refining operations.

Naturally, unsaturated fatty acids in vegetable oils are in the *cis* configuration, but during partial hydrogenation or as a result of thermal industrial (refining, deodorization and bleaching) or domestic processes (frying operations), *cis* fatty acids may be isomerized to the *trans* configuration [51-53]. Gas chromatography (GC) and Fourier transform infrared spectroscopy (FT-IR) are the official methods used for the determination of *trans* fatty acids in edible oils and fats [54]. FT-IR spectroscopy is a simple and rapid technique for the determination of *trans* fatty acids allowing the quantification of isolated *trans* C=C double bonds by measuring the peak area in the region 990-945 cm^{-1}, assigned to the C-H out-of-plane deformation corresponding to the *trans* configuration of the double bond. Recently, the accumulation of *trans* fatty acids in four types of Romanian vegetable oils (sunflower, soybean, corn and linseed) during thermal processing was studied [55]. Oils were continuously heated at three different temperatures (180, 220 and 250°C) for 33 h without any replenishment and the formation of the *trans* isomers was correlated with temperature, time and the unsaturation degree of oils. The results were compared with those obtained by gas chromatography and similar total *trans* fatty acids levels were found.

Transmission mode FT-IR spectroscopy was used to determine the amount of *trans* fatty acids in fourteen industrially hydrogenated and deodorized oils [51]. Better sensitivity of the method was achieved when a 200 μm KCl cell was used in transmission mode. The results of transmission FT-IR spectroscopy were evaluated by gas chromatography (GC) with flame ionization detector (FID), and found to be comparable. *Trans* fatty acids contents of partially hydrogenated oil samples were relatively higher as compared to those of the cooking oils. Among the samples examined, the highest level was found to be at 26.5% and 25.7% as determined by GC–FID and FT-IR methods, respectively.

The impact of heated edible oils on the intake of *trans* fatty acids was assessed by determining the *trans* isomerization of fatty acids in simulated frying and heating in home cooking conditions by gas-chromatography [52]. The amount of *trans* fatty acids in both fried potatoes and in frying oils was measured by gas chromatography (GC). For the frying model, sliced raw potatoes (10% of the frying oil, w/w) were fried in commercially available canola oil at 160, 180 and 200°C, with 10 frying cycles. *Trans* fatty acids content, acid values and peroxide values of the frying oils were compared with those of the correspondingly heated simple canola oil (without potatoes), used as a reference. The amounts of elaidic acid (*trans* 18:1) contained both in the frying and in the heated oils were less than the quantitative limit (0.047 g/100 g oil). The increases of *trans* 18:2 and *trans* 18:3 isomers in the used frying oil were 0.02 g/100 and 0.05 g/100 g, respectively, compared with those of the fresh oil. The accumulation of *trans* 18:2 fatty acids in the heated canola oil was slightly smaller than that in the frying oil. To elucidate the formation of the *trans* isomers in various edible oils during cooking, six kinds of commercially available edible vegetable oils were heated to 180°C in glass test tubes. Small changes in *trans* fatty acids profiles were observed after four hours of heating, suggesting that an ordinary frying process using non-hydrogenated edible oils has little impact on *trans* fats intake from edible oils.

The advantages of the FT-IR method are related to the fact that it may be applied directly on the oil sample without any pretreatment, thus being fast and non-destructive; the low price of the equipment is also acknowledged. The main drawbacks are related with the relatively low sensitivity and the fact that the method allows the quantification of the total *trans* isomers, not the identification and quantification of individual *trans* fatty acids [53].

On the other hand, chromatographic methods (either GC coupled with mass spectrometry, or with peak identification and quantification based on a standard *trans* FAME mixture), although being tedious compared to FT-IR (sample derivatization to the corresponding fatty acid methyl esters is prerequisite), allow both identification and quantification of individual *trans* fatty acids [51].

The presence of *trans* fatty acids can be detected by ^{1}H-NMR spectroscopy only at magnetic field strengths higher than 600 MHz on the basis of the signals corresponding to allylic protons in the 1.9-2.2 ppm spectral region [56]. The signals assigned to allyl groups adjacent to *cis* and *trans* isomers cannot be separated without sample derivatization at lower fields, being partially overlapped. Performant instruments operating at frequencies higher than 600 MHz allow full signal separation without prerequisite sample preparation. At field strengths as low as 300 MHz, the signal overlapping can be resolved on the basis of different complexation strengths of silver ions with *cis* and *trans* double bonds. The chelating agent is AgFOD (FOD = 4,4,5,5,6,6,6-heptafluoro-1-(2-thienyl)-1,3-hexanedienone) and the chelate: substrate molar ratio is 0.5. The induced separation of the allylic signals is similar to that obtained at 600 MHz (*Figure 3.3*)

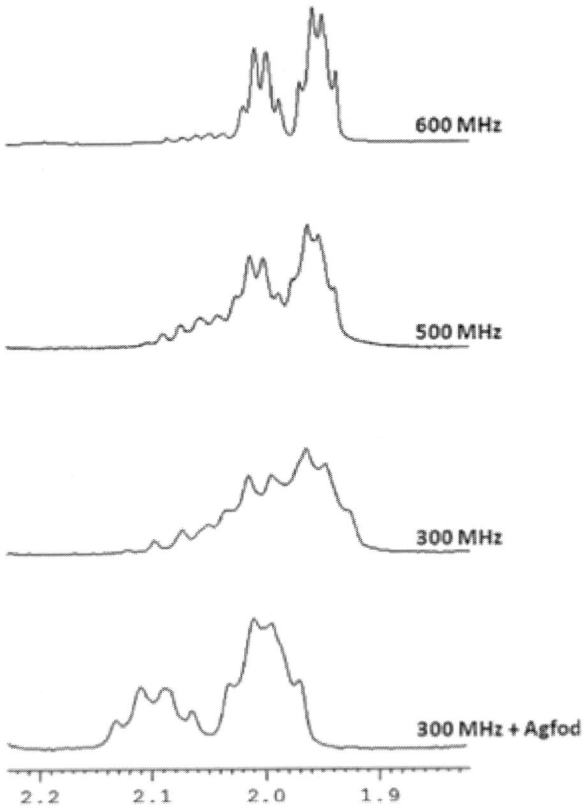

Figure 3.3. ^{1}H-NMR signals of the allylic protons of a 0.2M mixture of methyl oleate and methyl elaidate in CDCl3 recorded at different Larmor frequencies. The bottom spectrum was recorded after the addition of 0.1M AgFOD (silver chelate: lipid ratio = 0.5).

In the case of olive oils, another type of fraud was popular in the early 1970s: chemically synthesized triacylglycerols (TAGs) obtained by esterification of the free fatty acids rezulted from the refining process of olive oils with glycerol were sold as genuine olive oil. The raw material was indeed olive oil and consequently the composition was similar to that of olive oils, however the resulted oil was made of chemically synthesized triacylglycerols, therefore, it was not genuine and selling it as a genuine olive oil was a fraud. If the adulteration of high value oils from specific botanical origin (e.g., olive oils, almond oil) with cheaper oils of different botanical origin (e.g., sunflower oil) is relatively simple to assess on the basis of the compositional changes, the adulteration of virgin olive oil with synthesized TAG was a challenging issue in the 1980s. The discrimination of natural TAG from chemically synthesized ones was performed by analysing the positional distribution of acyl chains on the glycerol backbone [57-59]. It was established that the biosynthesized triacylglycerols do not allow saturated fatty acids or fatty acids with more than 18 carbon atoms in the *sn*-2 position. In return, the chemical synthesis leads to a random distribution of all types of fatty acids (both saturated and unsaturated) on the *sn*-1, -2 and -3 positions of the resulting TAG. Therefore, the analytical approaches focused on the direct analysis of the lipolysis reaction (with lipases specific for the *sn*-2 position) and the analysis of the resulting free fatty acids by gas chromatography. The method was tedious and time-consuming, however sensitive given the accuracy of the GC method. The method was improved in 1985 by Lerker et al. [60] by removing the TLC separation step, rendering it both easier and faster. After being validated, the method was officially adopted in 2002 and included in the Italian Regulation [61]. More recently, in the 1999, the ^{13}C-NMR spectroscopy allowed the rapid identification of saturated fatty acids esterified at the *sn*-2 position of triacylglycerols, after complete assignment of the peaks in the carbonyl region of the spectra based on model compounds [58]. The ^{13}C-NMR spectra of oils display in the 170-180 ppm region two groups of signals, corresponding to *sn*-1(3) and *sn*-2 positions, respectively. A genuine olive oil may display signals corresponding to stearic or palmitic moieties only in the deshielded group, assigned to *sn*-1(3) position, but the presence of signals from saturated fatty acids within the second group of signals will unambiguousely indicate an oil obtained through chemical synthesis. The advantage of the method is mainly due to its rapidity, however, the GC-based methods are more sensitive.

3.6. Detection of Adulteration

There is a common practice to adulterate expensive oils and fats (such as virgin olive oils or cocoa butter) with cheaper oils, for example with refined seed oils. The EU legislation (EU Regulation 1169/2011, effective since 2014) introduced a number of changes regarding food labelling. In the case of blended vegetable oils used in food products, the type of oil should be clearly indicated on the label, in contrast to the previous situation where an oil blend could have been generically specified as "vegetable oil" [45].

Evaluation of the linolenic acid content and the oleic/linoleic acids ratio allowed the detection of olive oil adulteration with refined seed oils [62]. Another simple and rapid indicator for establishing whether oils labelled as "virgin" contain refined oils is the specific extinction (SE) [63]. The SE of an oil will increase on refining because the usual 1:4 methylene interrupted distribution of double bonds found in linoleic and linolenic acids is altered in part to form a conjugated 1:3 distribution. If the oil has a SE greater than 0.25 at

270 nm, it is considered either to be not virgin or to contain oxidized fatty acids. If an oil is considered suspect, it is treated with alumina, which removes oxidation products, and the SE is determined again. If, after treatment with alumina, the SE is greater than 0.10, the oil is considered to be adulterated with refined oils.

The presence of *trans* isomers of oleic, linoleic or linolenic acids in the fatty acids profile of cold pressed olive oil above its characteristic limits may indicate the adulteration with either hydrogenated oils [64], refined olive oils or even with seed oils (e.g., sunflower) [65].

The detection of refined (deodorized or neutralized) vegetable oils as adulterants for olive oils may be carried out through quantification of the positional isomers of linoleic acid methyl esters [66], as well as by the evaluation of the diglyceride content, respectively; if the diglyceride content is not equal to the free fatty acids content, then the free fatty acids have been previously removed during the neutralization stage of the refining process, which may be evidence of adulteration with a refined oil [67]. Through refining (during bleaching) sterols are partially removed and the presence of their dehydration products (Δ3, 5-steradienes) into oils declared as un-refined (e.g., virgin olive oils, cold-pressed oils) is a good indicator for either adulteration with refined oils, or even for the fact that the investigated oil was refined [68, 69]. Even if sterols may be completely removed through advanced refining, their degradation products still remain in the refined oils [70].

The analysis of sterols is a particularly useful approach to identify adulteration of virgin olive oils with liquid vegetable oils, which tend to contain considerably higher levels of desmethylsterols than olive oils. Important chromatographic developments (including performant stationary phases) now allow the separation of 16 desmethylsterols from olive oils. It has internationally been agreed that the limit of "apparent β-sitosterol" (referring to the sum of the concentrations of β-sitosterol, Δ-5 avenasterol, Δ-5,23-stigmastadienol, Δ-5,24-stigmastadienol, cholesterol and sitostanol) should be greater than 93% of the total sterols [63].

Another example of the utility of sterol analysis was the detection of the presence of animal fats as adulterants for expensive vegetable solid or semi-solid fats (such as cocoa butter) and *vice versa*, on the basis of the evaluation of cholesterol (of animal origin) and fitosterols (of vegetable origin) content [71].

An original computational approach to predict proportions of different co-occurring oil varieties from quantitative chemical composition of blends was recently developed, consisting in applying a complete set of *N* mixtures of different olive oil varieties by gradually varying their proportions. The *N* simulated mixtures were characterized by *N* average fatty acid compositions computed from *N* combinations of randomly sampled individual fatty acids profiles. After *k* iterations of the mixture design, the *k* sets of *N* fatty acids average profiles were used as input data in a discriminant analysis to predict proportions of co-occurring olive-oil varieties in different blends. The predictive model was validated on outside blends, showing prediction errors with an order of 10% susceptible of reduction by applying a larger mixture design [40].

Conclusion

The authentication and quality control of vegetable oils are very important issues for the food industry. The demand of consumers for high quality oils or for oils with specific regional and varietal characteristics (PDO labels in the case of olive oils) and the threat of counterfeited products have led to the development of modern analytical techniques able to detect small changes (composition or even spectral characteristics) leading to prediction models to assess authenticity and the detection of adulteration.

Among the most intensively used analytical methods, NMR and vibrational (FT-IR, FT-Raman and NIR) spectroscopies proved their efficiency as screening techniques in order to acquire raw data which are further processed by chemometrical means.

These authentication methods are continuously improved by more and more sensitive instruments of analysis and also by the use of new techniques for the data processing. The improvement of authentication and detection of adulteration methods in the vegetable oils field would be facilitated by a closer collaboration between researchers and industrials, combining the scientific knowledge and the expensive infrastructure with the experience in the vegetable oils production.

REFERENCES

[1] Nunes, C. A. (2014). Vibrational spectroscopy and chemometrics to assess authenticity, adulteration and intrinsec quality parameters of edible oils and fats. *Food Res. Int.*, *60*, 255-261.

[2] Moore, J. C., Spink, J. & Lipp, M. (2012). Development and application of a database of food ingredient fraud and economically motivated adulteration from 1980 to 2010. *J. Food Sci.*, *77*, R118-R126.

[3] Ali, K., Verpoorte, R., Wilson, E. G. & Choi, Y. H. (2015). "Chemical Characterization of Marine Organisms Using a Metabolomic Approach" in "Marine Biomedicine. From Beach to Bedside." Baker, B.J. (Editor), CRC Press, Taylor & Francis Group, Boca Raton.

[4] van de Voort, F. R., Sedman, J. & Russin, T. (2001). Lipid analysis by vibrational spectroscopy. *Eur. J. Lipid Sci. Technol.*, *103*, 815-826.

[5] Larkin, B. (2011). Infrared and Raman spectroscopy, principles and spectral interpretation. Elsevier, Amsterdam.

[6] Stuart, B. H., (2004). Infrared Spectroscopy: Fundamentals and Applications, John Wiley & Sons, Chichester.

[7] Baeten, V., Hourant, P., Morales, M. T. & Aparicio, R. (1998). Edible oils and fats authentication by Fourier transform Raman spectrometry. *J. Agric. Food Chem.*, *46*, 2638-2646.

[8] Dijkstra, A. J., Christie, W. W. & Knothe, G. (2007). "Nuclear Magnetic Resonance Spectroscopy," in "*The Lipid Handbook*," 3rd Edition, Ed. CRC Press, New York, 455-467.

[9] Deleanu, C. (2005). "Nuclear Magnetic Resonance Spectroscopy Applications. Food," in "Encyclopedia of Analytical Science," Volume *6*, 2nd Edition, Worsfold, P. J., Townshend, A., Poole, C. F. (Editors), Elsevier, Oxford.

[10] Brescia, M. A. & Sacco, A. (2006). "High-Resolution 13C Nuclear Magnetic Resonance in the Study of Oils" in "Modern Magnetic Resonance," Graham A. Webb (Editor), Springer, New York.

[11] Chira, N. A., Nicolescu, A., Stan, R. & Rosca, S. (2016). Fatty Acid Composition of Vegetable Oils Determined From ^{13}C-NMR Spectra. *Rev. Chim.*, *67*(7), xxx-xxx – in press.

[12] Manina, L. & Segre, A. (2002). High resolution nuclear magnetic resonance: from chemical structure to food authenticity *Grasas y aceytes*, *53*, 22-33.

[13] Friebolin, H. (1991). "Basic One- and Two Dimensional NMR Spectroscopy," VCH, Weinheim.

[14] Yong, L. & Watkins, B. A. (2001). "Analysis of Fatty Acids in Food Lipids," in "Current Protocols in Food Analytical Chemistry," John Wiley & Sons, Inc., D.1.1.-D.1.2.

[15] Skorepa, J., Kahudova, V., Kotr, E., Mare, P. & Todorov, H. (1983). Gas-liquid chromatography profiling of intact lipids: Observation of differences between triglyceride structure of lipoproteins in type III and type IV hyperlipoproteinemia. *J. Chromatogr.*, *B. 273*, 180-186.

[16] Hoving, E. B., Jansen, G., Volmer, M., van Doormaal, J. J. & Muskiet, F. A. J. (1988). Profiling of plasma cholesterol ester and triglyceride fatty acids as their methyl esters by capillary gas chromatography, preceded by a rapid aminopropyl-silica column chromatographic separation of lipid classes. *J. Chromatogr.*, *434*, 395-401.

[17] Chira, N. A., Todaşcă, M. C., Nicolescu, A., Roşu, A., Nicolae, M. & Roşca, S. I. (2011). Evaluation of the computational methods for determining vegetable oils composition using ^{1}H NMR spectroscopy. *Rev. Chim.*, *62(1)*, 42-46.

[18] Barison, A., Werner Pereira da Silva, C., Ramos Campos, F., Simonelli, F., Antonio Lenz, C. & Gilberto Ferreira, A. (2010). A simple methodology for the determination of fatty acids composition in edible oils through ^{1}H-NMR spectroscopy. *Magn. Reson. Chem.*, *48(8)*, 642-650.

[19] Esbensen, K. H., Guyot, D., Westad, F. & Houmoller, L. P. (2002). Multivariate Data Analysis - in Practice: an Introduction to Multivariate Data Analysis and Experimental Design, Camo Software AS., Oslo, Norway.

[20] Gemperline, P. J. (2006). "Principal Components Analysis" in "Practical Guide Chemometrics," 2nd Edition, Gemperline, P.J. (editor), CRC Press Taylor & Francis, Boca Raton.

[21] Chira, N. A., Todaşcă, M. C., Păunescu, G., David, I., Ionescu, N., Stanciu, M. & Roşca, S. (2011). Romanian Vegetable Oils Authentication by Multivariate Analysis of ^{1}H-NMR Data. *Rev. Chim.*, *62*(4), 391-395.

[22] Burkard, U. (2003). "Methods for Data Analysis" in "Chemoinformatics: a Textbook," Gasteiger, J., Engel, T. (Editors), Wiley-VCH Verlag GmbH&Co., Darmstadt.

[23] Vigli, G., Philippidis, A., Spyros, A. & Dais, P. (2003). Classification of Edible Oils by Employing ^{31}P and ^{1}H NMR Spectroscopy in Combination with Multivariate Statistical Analysis. A Proposal for the Detection of Seed Oil Adulteration in Virgin Olive Oils. *J. Agric. Food Chem.*, *51*, 5715-5722.

[24] Meier, P. C. & Zund, R. E. (2000). Statistical Methods in Analytical Chemistry, 2nd Edition, John Wiley and Sons, Inc., New York.

[25] Leach, A. R. & Gillet, V. J. (2007). "An Introduction to Chemoinformatics. Revised Edition," Springer, Dordrecht.

[26] Dais, P. & Hatzakis, E. (2013). Quality assessment and authentication of virgin olive oil by NMR spectroscopy: A critical review. *Anal. Chim. Acta*, *765*, 1-27.

[27] Gomez-Caravaca, A. M., Maggio, R. M. & Cerretani, L. (2016). Chemometric applications to assess quality and critical parameters of Virgin and Extra-Virgin Olive Oil. A Review. *Anal. Chim. Acta*, in press. DOI: 10.1016/j.aca.2016.01.025.

[28] Aparicio, R., Morales, M. T., Aparicio-Ruiz, R., Tena, N., Garcia-Gonzalez. (2013). Authenticity of olive oil: Mapping and comparing official methods and promising alternatives. *Food Res. Int.*, *54*, 2025-2038,

[29] Marighedo, N. A., Kemsley, E. K., Defernez, M. & Wilson, R. H. (1998). A comparison of mid-infrared and Raman spectroscopies for the authentication of edible oils. *J. Am. Oil Chem. Soc.*, *75*, 987-992.

[30] Yang, H., Irudayaraj, J. & Paradkar, M. M. (2005). Discriminant analysis of edible oils and fats by FTIR, FT-NIR and FT-Raman spectroscopy. *Food Chem.*, *93*, 25-32.

[31] Javdinia, K., Parish, M., Karimi, S. & Hemmateenejad, B. (2013). Discrimination of edible oils and fats by combination of multivariate pattern recognition and FT-IR spectroscopy: A comparative study between different modeling methods. *Spectrochimica Acta Part A: Molecular and Biomolecular Spectroscopy*, *104*, 175-181.

[32] Mihalache, M., Bratu, A., Hanganu, A., Chira, N. A., Todasca, M. C. & Rosca, S. (2012). A New Chemometric Strategy Based on ^{1}H-NMR Data Applied for Authentication of Romanian Vegetable Oils. *Rev. Chim.*, *63(9)*, 877-882.

[33] Hanganu, A., Todasca, M. C., Chira, N. A., Maganu, M. & Rosca, S. (2012). The compositional characterisation of Romanian grape seed oils using spectroscopic methods. *Food Chem.*, *134*(4), 2453-2458.

[34] De Luca, M., Restuccia, D., Clodoveo, M. L., Puoci, F. & Ragno, G. (2006). Chemometric analysis for discrimination of extra virgin olive oils from whole and stoned olive pastes. *Food Chem.*, *202*, 432-437.

[35] Amirante, P., Clodoveo, M. L., Dugo, G., Leone, A. & Tamborrino, A. (2006). Advance technology in virgin olive oil production from traditional and de-stoned pastes: Influence of the introduction of a heat exchanger on oil quality. *Food Chem.*, *98*, 797–805.

[36] Clodoveo, M. L., Hbaieb, R. H., Kotti, F., Mugnozza, G. S. & Gargouri M. (2014). Mechanical strategies to increase nutritional and sensory quality of virgin olive oil by modulating the endogenous enzyme activities *Comprehensive Reviews in Food Science and Food Safety*, *13*, 135–154.

[37] Restuccia, D., Spizzirri, U. G., Chiricosta, S., Puoci, F., Altimari, I. & Picci, N. (2011). Antioxidant properties of EVOO from Cerasuola cv. olive fruit: Effect of stone removal. *Ital. J. Food Sci.*, *23*, 62–71.

[38] Servili, M., Taticchi, A., Esposto, S., Urbani, S., Selvaggini, R. & Montedoro, G. (2007). Effect of olive stoning on the volatile and phenolic composition of virgin olive oil. *J. Agric. Food Chem.*, *55*, 7028–7035.

[39] del Rio, J. C., Evaristo, A. B., Marques, G., Martin-Ramos, P., Martin-Gil, J. & Gutierrez, A. (2016). Chemical composition and thermal behavior of the pulp and kernel oils from macauba palm (*Acrocomia aculeata*) fruit. *Ind. Crops Prod.*, *84*, 294-302.

[40] Semmar, N. & Artaud, J. (2015). A new simplex-based approach predicting olive oil blend compositions from fatty acid data. *J. Food Comp. Anal.*, *43*, 149-159.

[41] Mannina, L., Marini, F., Gobbino, M., Sobolev, A. P. & Capitani, D. (2010). NMR and Chemometrics in Tracing European Olive Oils: The Case Study of Ligurian Samples. *Talanta, 80*, 2141-2148.

[42] Mannina, L., Fontanazza, G., Patumi, M., Ansanelli, G. & Segre, A. L. (2001). Italian and Argentine olive oils: A NMR and gas-chromatographic study. *Grasa y Aceites, 6*, 380-388.

[43] Alonso-Salces, R. M., Moreno-Rojas, J. M., Holland, M. V., Reniero, F., Guillou, C. & Herberger, K. (2010). Virgin olive oil authentication by multivariate analyses of ^{1}H NMR fingerprints and δ^{13}C and δ^2H data. *J. Agric. Food Chem., 56*, 5586-5596.

[44] Rezzi, S., Axelson, D. E., Heberger, K., Reniero, F., Mariani, C. & Guillou, C. (2005). Classification of olive oils using high throughput flow ^{1}H NMR fingerprinting with principal component analysis, linear discriminant analysis and probabilistic neural networks. *Anal. Chim. Acta, 552*, 13-24.

[45] Osorio, M. T., Haughey, S. A., Elliott, C. T. & Koidis, A. (2014). Evaluation of methodologies to determine vegetable oils species present in oil mixtures: Proposition of an approach to meet the EU legislation demands for correct vegetable oils labelling. *Food Res. Int., 60*, 66-75.

[46] Locatelli, M., Coisson, J. D., Travaglia, F., Cereti, E., Garino, C., D'Andrea, M., Martelli, A. & Alorio, M. (2011). Chemotype and genotype chemometrical evaluation applied to authentication and traceability of "Tonda Gentile Trilobata" hazelnuts from Piedmont (Italy). *Food Chem., 129*, 1865-1873.

[47] Agiomyrgianaki, A., Petrakis, P. V. & Dais, P. (2012). Influence of harvest year, cultivar and geographical origin on Greek extra virgin olive oils composition: A study by NMR spectroscopy and biometric analysis. *Food Chem., 135*, 2561-2568.

[48] Berrueta, L. A., Alonso-Salces, R. M. & Heberger, K. (2007). Supervised pattern recognition in food analysis. *J. Chromatogr., A 1158*, 196-214.

[49] D'Imperio, M., Gobbino, M., Picanza, A., Costanzo, S., Della Corte, A. & Mannina, L. (2010). Influence of harvest method and period on olive oil composition: an NMR and statistical study, *J. Agric. Food Chem., 58*, 11043–11051.

[50] Beltrán, G., del Rio, C., Sánchez, S. & Martínez, L. (2004). Influence of Harvest Date and Crop Yield on the Fatty Acid Composition of Virgin Olive Oils from Cv. Picual. *J. Agric. Food Chem., 52* (11), 3434–3440.

[51] Sherazi, S., Kandhro, A. & Arain, S. (2009). Application of transmission FT-IR spectroscopy for the *trans* fat determination in the industrially processed edible oils. *Food Chem., 114*(1), 323-327.

[52] Tsuzuki, W., Matsuoka, A. & Ushida, K. (2010). Formation of *trans* fatty acids in edible oils during the frying and heating process. *Food Chem., 123*, 976-982.

[53] Cho, I. K., Kim, S. K., Khurana, H. K., Li, Q. X. & Jun, S. (2011). Quantification of *trans* fatty acids content in French fries of local food service retailers using attenuated total reflection- Fourier Transform infrared spectroscopy. *Food Chem., 125*, 1121-1125.

[54] Priego-Capote, F., Ruiz-Jimenez, J. & Luque de Castro, M. D. (2007). Identification and quantification of *trans* fatty acids in bakery products by gas-chromatography-mass spectrometry after focused microwave Soxhlet extraction. *Food Chem., 100*, 859-867.

[55] Mihalache, M., Bratu, A., Hanganu, A., Chira, N. A., Maganu, M., Todasca, M. C. & Rosca, S. (2012). Thermal Formation of *trans* Fatty acids in Romanian Vegetable Oils Monitored by GC-MS and FT-IR Techniques. *Rev. Chim., 63(10)*, 984-988.

[56] Agiomyrgianaki, A., Sedman, J., van der Voort, F. R. & Dais, P. (2012). *Cis* and *trans* components of lipids: Analysis by ^{1}H-NMR and silver shift reagents. *Eur. J. Lipid Sci. Technol.*, *114*(5), 504-509.

[57] Sacchi, R., Adeo, F. & Paolillo, L. (1997). ^{1}H and ^{13}C NMR of olive oil. An overview. *Magn. Reson. Chem.*, *35*, S133-S145.

[58] Vlahov, G. (1999). Application of NMR to the study of olive oils. *Progr. NMR. Spectrosc.*, *35(4)*, 341-357.

[59] Mannina, L., Luchinat, C., Emanuele, M. C. & Segre, A. L. (1999). Acyl positional distribution of glycerol tri-esters in vegetable oils: A ^{13}C NMR study. *Chem. Phys. Lipids*, *103*, 47-55.

[60] Lerker, G., Moschetta, M., Caboni, M. & Frega, N. (1985). Determinazione degli oli esterificati negli oli provenienti dalla lavorazione delle olive. *Rivista Italiana delle Sostanze Grasse*, *62*, 15-18.

[61] Norme Grassi e Derivati, (2002). NGD C 86-03. Determinazione della percentuale di acido palmitico in posizione 2 dei trigliceridi mediante dosaggio del 2 gliceril monopalmitato. Stazione Sperimentale per le Industrie degli Oli e dei Grassi (SSOG), Milano, Italy.

[62] Kapoulas, V. M. & Passaloglou-Emmanouilidou, S. (1981). Detection of adulteration of olive oil with seed oils by a combination of column and gas liquid chromatography. *J. Am. Oil. Chem. Soc.*, *58*, 694-702.

[63] Bell, J. R. & Gillatt, P. N. (2016). Standards to ensure the authenticity of edible oils and fats. FAO Corporate Document Repository, http://www.fao.org/docrep/t4660t/t4660t0e.htm, accessed on Feb. 18th.

[64] Grob, K., Giuffre, A. M., Leuzzi, U. & Mincione, B. (1994). Recognition of adulterated oils by direct analysis of the minor components. *Fat Sci. Technol.*, *96*, 286-290.

[65] Skevin, D., Kraljic, K., Miletic, L., Obranovic, M., Nederal, S. & Petricevic, S. (2011). Adulteration of Oblica Virgin Olive Oil with Edible Sunflower and Refined Olive Pomace Oil. *Croat. J. Food Technol. Biotechnol. Nutr.*, *6*(3-4), 117-122.

[66] European Communities (EC), (1992). *Official Journal of the Commission of the European Communities.* Regulation No. 1429/92, L150, May 26th, 1992.

[67] Capella, P. (1993). Analisis del aceite de oliva. *Olivae*, *45*(2), 24-27.

[68] Leon-Camacho, M., Alvarez Serrano, M. & Graciani Constante, E. (2004). Formation of stigmasta-3, 5-diene in olive oil during deodorization and/or physical refining using nitrogen as stripping gas. *Grasas y Aceitas*, *55*(3), 227-232.

[69] Cert, A., Lanzon, A., Carelli, A. A., Albi, T. & Amelotti, G. (1994). Formation of stigmas-3, 5-diene in vegetable oils. *Food Chem.*, *49*, 287-293.

[70] Ulberth, F. & Buchgraber, M. (2000). Authenticity of fats and oils. *Eur. J. Lipid Sci. Technol.*, *102*, 687-694.

[71] Homberg, E. (1991). Sterol analysis as means to detection of mixed and adulterated fats. *Fat Sci. Technol.*, *93*, 516-517.

In: Vegetable Oil
Editor: Brittany Holt

ISBN: 978-1-63485-128-2

Chapter 3

BIODIESEL PRODUCTION FROM VEGETABLE OILS: A SUSTAINABLE ENERGY ALTERNATIVE

Edith Martinez-Guerra and Veera Gnaneswar Gude*
Department of Civil and Environmental Engineering
Mississippi State University, Mississippi State, MS, US

ABSTRACT

Declining global fossil fuel reserves due to escalating consumption and associated environmental pollution created an urge for investigating into production of renewable and environmentally friendly biofuels. Biofuels such as biodiesel are most sought because they can be produced in liquid form suitable for most transportation needs. Moreover, biodiesel can be produced from a variety of oil-rich feedstock and even from waste cooking oils, animal fat and other microbial lipids. This book chapter provides an overview of the oil feedstock suitable for biodiesel production and their characteristics. The chemistry and process of biodiesel production are discussed in detail.

Keywords: biodiesel, esterification, microwave, transesterification, ultrasound, vegetable oil, waste cooking oil

INTRODUCTION

Biodiesel is a biodegradable and renewable fuel which can be produced from a variety of feedstock and is currently being produced in many parts of the world. It is considered the fuel of the future and could present as a major competitor for petroleum diesel in near future. Biodiesel can be used by itself in existing diesel engines without any modification to the engine, or blended at any ratio with petroleum diesel as it is currently practiced worldwide [1-2]. Biodiesel production may not completely replace the fossil fuel consumption, rather it can aid to diminish the dependency on the conventional fossil fuel sources. One of the major

* E-mail: gude@cee.msstate.edu; gudevg@gmail.com, Ph. 662-325-0345.

drawbacks of biodiesel is the feedstock availability for its production, and energy-, chemical-, and cost-efficient process methods that can make biodiesel cost-competitive with petroleum diesel. The concern for the lack of this process method has motivated and challenged researchers to seek for highly efficient technique that can produce biodiesel and save time and energy simultaneously. Hitherto, conventional heating technologies have provided the means to convert lipids into fatty acid alkyl esters (FAAE). However, these technologies lack the capabilities required to make biodiesel production feasible at large-scale. Therefore, the need for non-conventional technologies is imminent. Thus, microwave and ultrasound irradiations have become the most prominent non-conventional technologies promising to revolutionize biodiesel production. Biodiesel can be produced from any raw material containing triglycerides or lipids in it, including algal oils, waste cooking oils, animal fats, and vegetable oils. Algal oils and waste cooking oils are two commonly used feedstocks for biodiesel production, which were both evaluated using microwave and/or ultrasound irradiations extensively to produce biodiesel. The use of vegetable oils both used (recycled) and unused as a feedstock for biodiesel production is discussed in detail in this book chapter. Although the selection of the most suitable feedstock is vital for biodiesel production, there are several more parameters that must be taken into account when selecting the most efficient process to produce biodiesel. These include the use of an effective solvent or reactant, a fast-reacting catalyst, and an optimum reaction time and process intensification techniques that result in energy, chemical and cost savings.

Biodiesel Feedstock

Biodiesel can be produced from a variety of oil-rich feedstock. The feedstock can be classified as conventional and non-conventional feedstock as shown in Table 1. Conventional feedstock represents the feedstock that was used in earlier biodiesel production research studies. These are soybean, peanut, canola, sunflower, corn and others. These feedstock were considered as first generation feedstock. The demand for these feedstock, however, has raised concerns or dilemmas known as *"Fuel vs. Food."* To avoid this issue, research on biodiesel production using non-conventional feedstock has been aggressively conducted throughout the world. These feedstock include animal fats, used oils such as waste cooking oils, grease and tallow and among others. These can be considered as second generation biodiesel feedstock. Finally, high oil or lipid yielding feedstock such as algae. Bacteria and other microbial species were considered for biodiesel production and these were being studied extensively in recent years.

The oil composition of feedstock varies from one origin to the other. Table 2 shows the major fatty acid composition (also known as triglycerides) for different oil feedstock. The most commonly found fatty acids are long chain C16, and C18 compounds. Examples of common fatty acids are stearic, oleic, linolenic and palmitic. Properties such as viscosity, specific density, flash point, heating value, acid value, cetane number, cloudpoint and pour point are shown in Table 2. The viscosity of vegetable oils is around 11–17 times more than diesel fuel. The volumetric heating values are around 39–40 MJ/kg but for diesel fuels, it is 45 MJ/kg. The flash point for vegetable oils is very high, more than 200°C. Vegetable oils can be directly used in engines but they cause clogging and smoke and precipitation issues.

These oils (fatty acids) need to be converted into biodiesel (alkyl esters) to address these issues. As a result, biodiesel has significant influences in reducing engine emissions such as unburned hydrocarbons (68%), particulars (40%), carbon monoxide (44%), sulfur oxide (100%), and polycyclic aromatic hydrocarbons (PAHs) (80–90%) [3].

Table 1. Different feedstocks for production of biodiesel [3]

Conventional feedstock	Non-conventional feedstock
Mahua, Soybean, Nile tilapia, Rapeseed, Palm, Canola, Poultry, Babassu, Tobacco seed, Brassica carinata, Rubber plant, Brassica napus, Rice bran, Copra, Sesame, Groundnut, Sunflower, Cynara cardunculus, Barley, Cottonseed, Coconut, Pumpkin, Corn, Jojoba oil, Used cooking oil Camelina, Linseed, Peanut, Mustard, Olive	Lard, Tallow, Poultry fat, Fish oil, Tarpenes, Latexes, Pongamina pinnata, Palanga, Jatropha curcas, Sea mango, Okra Bacteria, Fungi, Macro-algae, Micro-algae

Waste Cooking Oil (WCO) as a Fuel Source

Waste cooking oil is one of the most prominent feedstock for biodiesel production. Vegetable oils as a feedstock for biodiesel production were proposed in 1978 in the United States and 1981 in South Africa. Methyl ester was produced in Germany and Austria (1982) from rapeseed oil; however, it was not only until 1985 that a small pilot plant was built in Austria and commercial production began 15 years later in Europe [4]. The use of waste cooking oil for this purpose is not only economical and practical, but it also reduces environmental pollution. It is estimated that approximately 100 million gallons of WCO is produced per day in the United States, where 9 pounds of WCO are generated per person per year [5]. Therefore, the use WCO as a fuel source could greatly reduce the burden of its disposal. WCO can become a problem when disposed improperly, down the kitchen sink for example, where it can quickly cause blockages of sewer pipes when the oil solidifies; or properties of degraded WCO after it gets into sewage system are conductive to corrosion of metal and concrete element [6]. Moreover, if dumped in municipal solid waste landfills or into municipal sewage treatment plants, it creates operation problems along with water and soil pollution issues [7]. Previously, WCO was used to serve as animal feed; however, the European Union (EU) enforced a ban on WCO as animal feed because the harmful compounds present in WCO could return back into the food chain through animal meat [8]. It is also not advisable to recycle used cooking oils for edible purposes with high polar content greater than 25% [9].

As mentioned before, biodiesel can be produced by using several types of vegetable oils, which are often virgin oils; however, the cost of biodiesel produced from virgin vegetable oil through transesterification is higher than that of fossil fuel [10]. Also, WCO cost is two to three times cheaper than virgin oils, and it also reduces the cost of waste product removal and treatment [11]. With regard to, the conversion of waste cooking oil into biodiesel is rapid and simple, not energy-intensive extractive processes are necessary for the conversion.

Table 2. The properties of different vegetable oils [3, 4]

Type of Oil	Species	Major Fatty acid composition	Viscosity (at 40°C)	Density (g/cm3)	Flash point (°C)	Heating value (MJ/kg)	Acid value (mg KOH/g)	Cetane number (C)	Cloud point (°C)	Pour point (°C)
Edible oil	Soybean	C16:0, C18:1, C18:2	32.9	0.91	254	39.6	0.2	37.9	-3.9	-12.2
	Rapeseed	C16:0, C18:0, C18:1, C18:2	35.1	0.91	246	39.7	2.92	37.6	-3.9	-31.7
	Sunflower	C16:0, C18:0, C18:1, C18:2	32.6	0.92	274	39.6	–	41.3	18.3	-6.7
	Palm	C16:0, C18:0, C18:1, C18:2	39.6	0.92	267	–	0.1	42.0	31.0	–
	Peanut	C16:0, C18:0, C18:1, C18:2, C20:0, C22:0	22.72	0.90	271	39.8	3	41.8	12.8	_6.7
	Corn	C16:0, C18:0, C18:1, C18:2, C18:3	34.9	0.91	277	39.5	–	37.6	_1.1	_40.0
	Camelina	C16:0, C18:0, C18:1, C18:2, C18:3, C20:0, C20:1, C20:3	–	0.91	–	42.2	0.76	–	–	–
	Canola	C16:0, C18:0, C18:1, C18:2, C18:3	38.2				0.4	–	–	–
	Cotton	C16:0, C18:0, C18:1, C18:2	18.2	0.91	234	39.5		41.8	1.7	_15.0
	Pumpkin	C16:0, C18:0, C18:1, C18:2	35.6	0.92	>230	39	0.55	–	–	–
Non-edible oil	Jatropha curcas	C16:0, C16:1, C18:0, C18:1, C18:2	29.4	0.92	225	38.5	28	–	–	–
	Pongamina pinnata	C16:0, C18:0, C18:1, C18:2, C18:3	27.8	0.91	205	34	5.06	–	–	–
	Sea mango	C16:0, C18:0, C18:1, C18:2	29.6	0.92	–	40.86	0.24	–	–	–
	Palanga	C16:0, C18:0, C18:1, C18:2	72.0	0.90	221	39.25	44	–	–	–
	Tallow	C14:0, C16:0, C16:1, C17:0, C18:0, C18:1, C18:2	–	0.92	–	40.05		–	–	–
	Nile tilapia	C16:0, C18:1, C20:5, C22:6, other acids	32.1	0.91	–	–	2.81	–	–	–
	Poultry	C16:0, C16:1, C18:0, C18:1, C18:2, C18:3	–	0.90	–	39.4	–	–	–	–
Others	WCO	Depends on fresh cooking oil	44.7	0.90	–	–	2.5	–	–	–
–	Diesel	–	3.06	0.855	76	43.8	–	50	–	-16

According to Tomasevic and Siler-Marinkovic [12], the quantity of double bonds is lower in WCO, in comparison to virgin vegetable oil, and can produce biodiesel with a lower number of double bonds (improving iodine index) and, therefore, higher quality biodiesel. Despite all the advantages of using WCO for biodiesel production, there are also several disadvantages. For example, the quality of WCO depends on the amount of free fatty acids (FFA) and water content. WCO containing high amounts of FFA may require a pre-treatment process, or an extended transesterification process that will contribute to the total biodiesel cost. Additionally, oil degradation during cooking can occur. This usually happen through three main reactions: thermolytic, oxidative, and hydrolytic reactions. In thermolytic reactions, the reaction occurs in absence of oxygen. High temperature is required to decompose saturated fatty acids to form alkanes, fatty acids, ketones, esters, diacylglycerides. In addition, dimeric compounds appear to be the main derivatives as a result of thermolytic reactions of unsaturated fatty acids. The physical and chemical properties such as viscosity, water content, free fatty acid content, and consists of polymerized and oxidized compounds depend on the duration that the oil is exposed to food, heat, and oxygen during cooking [13].

WCO can be transesterified by using a base or an acid catalyst. These catalysts can be homogenous or heterogeneous, as explained in the catalyst section. Because of its low cost, a basic homogenous catalyst is often preferred for the conversion. However, during the transesterification reaction, the free fatty acids in WCO react with the base catalyst to form soap and water. The soap formation is also known as saponification. The water formation will hydrolyze triglyceride to form more free fatty acids [10]; thus, reducing the biodiesel quality. A major advantage of WCO base-catalyzed transesterification is that it requires only low temperatures and pressures, it involves direct conversion to biodiesel with no intermediate compounds, and no special materials of construction are needed [14]. The main byproduct formed during a base-catalyzed transesterification is glycerol, which is a useful product to produce soap, makeup, and other personal care products. Unfortunately, glycerol is difficult to recover from the transesterification and adds to the cost of biodiesel production. In order to overcome the soap formation, the homogenous basic catalyst is often replaced with an acid catalyst. According to Freedman et al. (1986), acid catalysts are insensitive to free fatty acids (FFA) and, therefore, they are more suitable for the conversion of WCO with high amounts of FFA [15]. Despite the benefits of acid-catalyzed transesterification (also known as esterification), the use of alkaline catalyst is predominantly because of acid catalyst slower reaction rate [16] and higher amounts are often required as well as higher temperatures and extended periods of reaction times. In consequence of hurdles of using either a base or an acid catalyst, a two-step transesterification is usually required. The two-step transesterification consist of esterifying of the FFA present in waste frying oils using an acid catalyst to decrease the FFAs to a level less than 1%. In the second step, transesterification of the obtained oils is performed using an alkaline catalyst [6].

TRANSESTERIFICATION

Transesterification is the conversion of triglycerides in oils to mono-alkyl esters (biodiesel) by using a monohydric alcohol and a catalyst [18]. The two major products of transesterification are esters (biodiesel) glycerol. During transesterification, triglycerides are sequentially (eq.1- eq. 3) converted to diglycerides, monoglycerides, and finally glycerol (by product), with the alkyl ester being produced at every step [19]. From equation 1: R1, R2, and R3 radicals represent long chain hrydrocarbon known as fatty acids. The transesterification process (Figure 1) reduces significantly the biodiesel production by-products resulting in higher quality and quantity FAAEs by facilitating the conversion of oil to biodiesel.

$$\boldsymbol{Triglyceride + R'OH \xleftrightarrow{Catalyst} Diglyceride + R'COOR_1} \quad (1)$$

$$\boldsymbol{Diglyceride + R'OH \leftrightarrow Monoglyceride + R'COOR_2} \quad (2)$$

$$\boldsymbol{Monoglyceride + R'OH \leftrightarrow Glycerol + R'COOR_3} \quad (3)$$

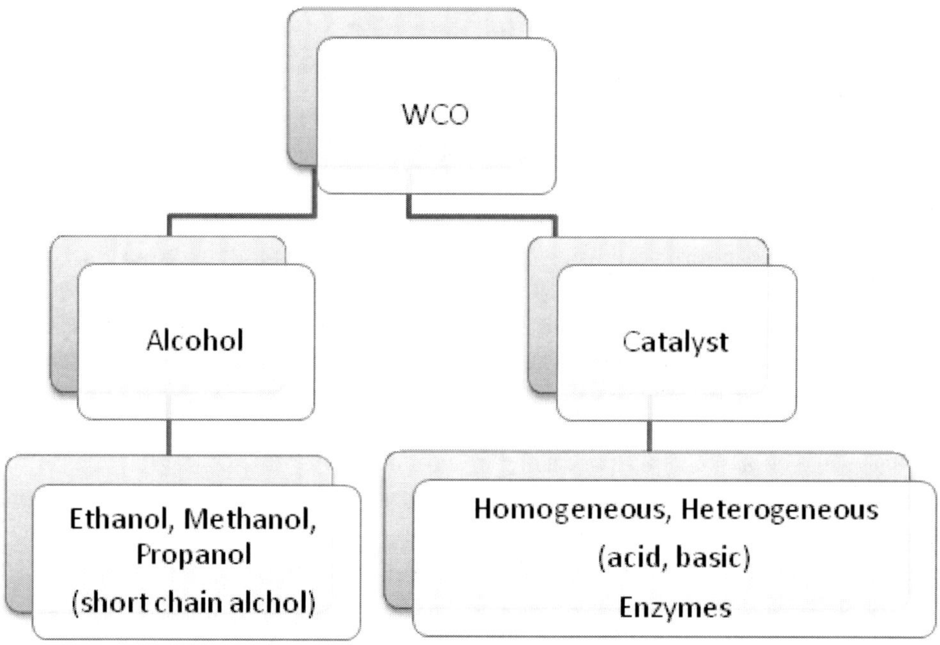

Figure 1. Critical parameters for transesterification reaction.

THE IMPORTANCE OF A CATALYST IN THE PRODUCTION OF BIODIESEL

Catalysts used for transesterification of biodiesel feedstock include homogeneous and heterogeneous chemical catalysts and enzymes other ion exchange catalysts. Homogeneous catalysts in general are very efficient with high reaction kinetics. Heterogeneous catalysts, either solid acids or solid bases are less corrosive than homogeneous catalysts but the reaction kinetics are almost 1000 times lower than homogeneous catalysts. However, better results can be obtained if the porosity of the heterogeneous catalyst is higher. Heterogeneous bases proceed by reaction of either the Lewis or the Brønsted basic sites of the catalyst with a monohydric alcohol, and the generated alkoxide mixture interacts with the esters in the oil to yield biodiesel [18].

Homogenous and Base Catalyst

Sodium hydroxide (NaOH) and potassium hydroxide (KOH) are the most commonly used homogenous base catalysts, and they are very efficient in achieving high conversion rates in short reaction time (2 min). However, the downside of alkali catalyzed process is saponification, the formation of undesirable amounts of soap. Another disadvantage of base catalysts is they are highly corrosive. The alkali mediated reactions consist of three steps (Figure 2): 1) the base reacts with the alcohol to produce an alkoxide and the protonated catalyst; 2) the nucleophilic attack of the alkoxide at the carbonyl group of the triglyceride generating a tetrahedral intermediate, from which the alkyl ester and the corresponding anion of the diglyceride are formed; and 3) the latter deprotonates the catalyst can react with second molecule of alcohol and start another catalytic cycle [20, 21]. Base catalyst is known to react faster than acids; however, acid catalysts are more efficient converting high FFA feedstock into biodiesel [22].

Homogeneous catalysts such as sodium hydroxide or potassium hydroxide dissolve in the biodiesel, and their separation process produces large quantities of wastewater impacting the environment, adding to the treatment processes and finally increasing product costs. These issues can be circumvented by employing heterogeneous catalysts that do not dissolve or sparingly dissolve in oil or methanol phases and eliminate the extensive cleaning, separation, and drying processes. Heterogeneous catalysis has many advantages such as non-corrosiveness, being an environmentally benign process, fewer disposal problems, and better biodiesel production economics [23]. They are also much easier to separate from liquid products and can be designed for higher activity, selectivity, and long catalyst lifetimes [24, 25]. Many types of heterogeneous catalysts, such as alkaline earth metal oxides, and various alkaline metal compounds supported on alumina or zeolite can catalyze transesterification reactions [26-28].

Base Heterogeneous Catalyst

The main mechanism of heterogeneous catalysis follows the principle similar to homogeneous catalysis of either acid or base systems [29, 30]. The important factor in homogeneous base catalyzed reaction is to create nucleophilic alkoxide from the alcohol to attack the electrophilic part of the carbonyl group of the triglycerides [31], while in acid catalysis the carbonyl group in triglycerides is protonated, and the alcohol attacks the protonated carbon to create a tetrahedral intermediate. The catalyst efficiency depends on several factors such as specific surface area, pore size, pore volume and active site concentration [32]. The order of activity among alkaline earth oxide catalysts is BaO > SrO > CaO > MgO [33-35]. The structure of metal oxides is made up of positive metal ions (cations) which possess Lewis acidity, i.e., they behave as electron acceptors, and negative oxygen ions (anions) which behave as proton acceptors and are thus Brønsted bases. This has consequences for adsorption. In methanolysis of oils, it provides sufficient adsorptive sites for methanol, in which the (O–H) bonds readily break into methoxide anions and hydrogen cations. The methoxide anions then react with triglyceride molecules to yield methyl esters [36, 37].

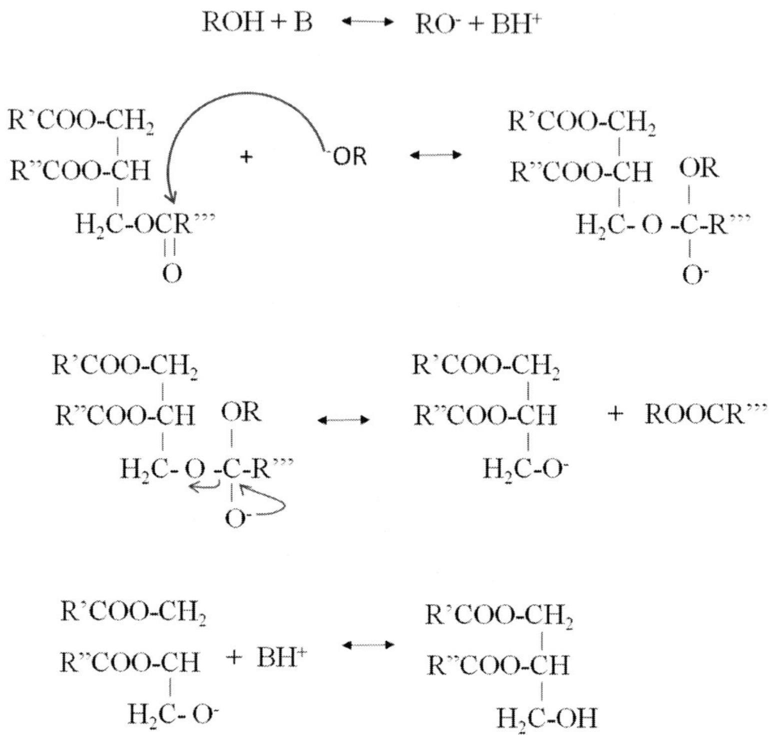

Figure 2. Base catalyst transesterification.

A common heterogeneous catalyst is barium oxide, and it catalyzes the transesterification reaction by forming barium methoxide with methanol. Due to its very low methanol solubility, barium methoxide acts mainly as a heterogeneous catalyst. Pure barium methoxide, which is strongly basic, shows high catalytic activity. It is not soluble in methanol but it forms a suspensoid, whereby its active surface is very well developed. Metal alkoxides can be homogeneous catalysts if they are well soluble in methanol or if they can constitute active centers on the surface of heterogeneous catalysts. The alkalinity of a given compound is a key factor, which determines its catalytic activity in alcoholysis. Alkaline-earth metal compounds are heterogeneous catalysts and the degree of their dispersion in the reaction system has a considerable influence on the level of their catalytic activity, which is determined by diffusion. The classic mechanism of base-catalyzed alcoholysis assumes an attack of the alkoxide anion on the carboxylate carbon of the ester group converting it to a tetrahedral intermediate carrying a negative charge. Methoxide anions can be introduced as alcoholates directly into the reaction if they are soluble in methanol and dissociate easily. The well-known catalytic activity of sodium methoxide in alcoholysis is a good example [38]. Methoxides can also result from reversible reactions of some metal oxides or hydroxides with methyl alcohol [39, 40]; e.g., the reaction of sodium hydroxide with methanol.

ACID CATALYST

The transesterification process by acids (Figure 3) gives high biodiesel yields and there is not soap formation during the process, but the reactions are slow. Some homogenous acid catalysts are H_2SO_4, HCL, BF_3, H_3PO_4, and some organic sulfonic acids [21]. The addition of the acid catalyst leads to protonation of the carbonyl group of the ester that results in carbocation which after nucleophilic attack of the alcohol produces the tetrahedral intermediate, which eliminates glycerol to form the new ester and regenerates the catalyst H^+ [21]. Nevertheless, acid catalyst required longer reactions time and higher catalyst amounts, which make it more expensive than basic catalysts. The conversion of lipids to biodiesel using an acid catalyst is also known as esterification.

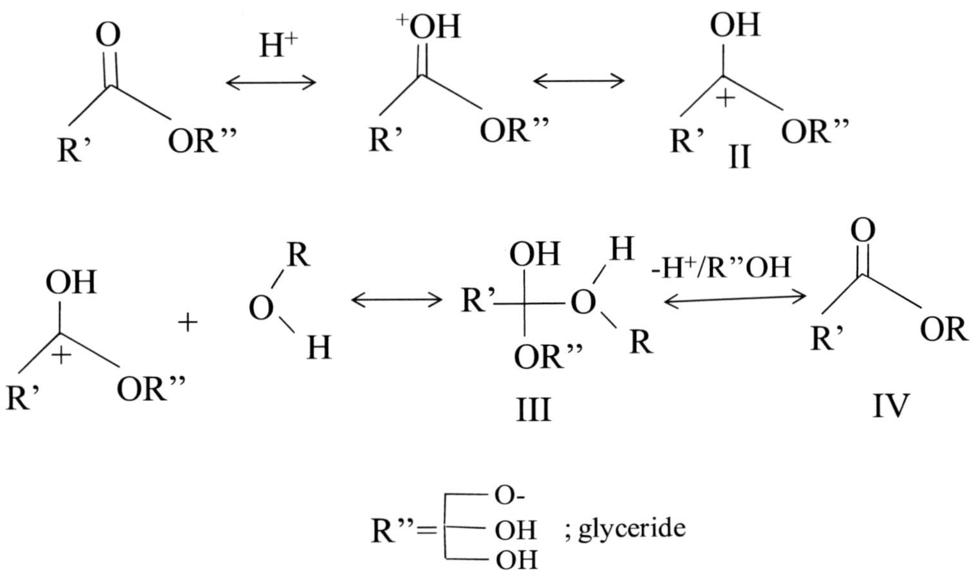

R'= carbon chain of the fatty acid
R= alkyl group of the alcohol

Figure 3. Acid-catalyzed transesterification.

OTHER ESSENTIAL PARAMETERS ON BIODIESEL PRODUCTION

In addition to the selection of a feasible feedstock and a fast reacting catalyst, biodiesel production requires of other vital parameters that play crucial roles in the optimization process. Some of these indispensable parameters include the selection of an adequate reactant/solvent, reaction time period, temperature, and power output of the selected technologies. The meticulous selection of these parameters determines the feasibility of biodiesel production.

Alcohol Selection and Oil to Alcohol Molar Ratio Effect

Transesterification reaction involves addition of an alcohol group form alkyl esters [41]. The short chain alcohols such as methanol, ethanol, and butanol are frequently used, with methanol being the most common. All of these alcohols possess different physical and chemical properties and therefore exhibit different patterns of transesterification reaction kinetics and biodiesel yield rates. Methanol and ethanol produce superior results in transesterification reactions. Currently, methanol is produced from the fossil sources and mineral oils [42] which impact the availability and cost of the oil resources. For long term sustainability of biodiesel production, use of ethanol provides a more environmentally friendly perspective since it can be derived from natural and renewable sources like plants and crops [42, 43]. Biodiesel containing ethyl esters are more beneficial compared to methyl esters because ethanol can be derived from renewable sources, and the extra carbon atom in the ethanol molecule slightly increases the heat content and the cetane number [44]. In addition, the ethyl esters generally have lower cloud and pour points than methyl esters [45].

Mass transfer issues are often not discussed in transesterification reactions. The transesterification reaction is mainly dominated by the mass transfer limitations, then followed by kinetics controlled region, with oils and methanol [46, 47]. Mass transfer between two organic phases (methanol and oil) plays a critical role during the transesterification (methanolysis) and controls the reaction kinetics. In transesterification reaction kinetics, three regimes are well-recognized, that is an initial mass transfer-controlled regime (slow), followed by a chemically-controlled regime (fast), and a final regime, close to equilibrium (slow) [48]. Therefore, methanol is not effectively used for the reactions due to the interfacial mass transfer resistance [49]. The mass transfer limitations on biodiesel production may be overcome with efficient mixing mechanism such as ultrasonication. Ultrasound can enhance the mass transfer between two immiscible liquids [50]. Cavitation mainly affects the mass transfer rates and ensures a uniform distribution of the reactants, as one concludes from the fact that a significant effect on both the reaction rate and the equilibrium conversion is only observed in the later stages of the reactions when heterogeneity is removed [51].

The transesterification reaction consists of three steps and, ideally, three moles of alcohol would be enough to carry out the conversion. Only one mole of alcohol would be consumed per each mole of ester produced. However, the reversible nature of the reaction demands excess of alcohol to move the reaction forward and to complete the formation of esters. Therefore, selecting the appropriate oil to alcohol molar ratio is very important for the transesterification process. Yet, there is still not an established optimum alcohol to oil molar ratio. For example, Pukale et al. (2015) reported that increasing the WCO to methanol molar ratio from 1:4 to 1:6, improved the biodiesel yield from 58 to 93% [52]. Leung and Guo (2015) obtained a biodiesel yield of 94% for a 1:7 WCO to alcohol ratio; however, with similar conditions Meng et al. (2008) obtained an 88.9% conversion yield [53, 54]. On the other hand, Dehkordi et al. (2012), experimented with higher WCO to alcohol molar ratios (1:3-1:60) resulting in biodiesel yields ranging from 22.7 to 92.1% [55]. Similarly, using higher molar ratios, Alves et al. (2013) reported an optimum WCO to alcohol molar ratio of 1:40 where they stated that higher molar ratios are particularly used when the triglyceride contains large amounts of free fatty acids [56]. Furthermore, Zheng et al. (2006) studied the effect of using oil to alcohol molar ratios in the ranges of 1:50-1:250, varying the amount of catalyst [57]. They observed that the conversion yield increased linearly with increasing the oil to alcohol molar ratio when using low amounts (1.5 mol%) of catalyst; however, they reached an optimum ratio of 1:50 when increasing the catalyst amount (2.5 mol%) that did not increase further at higher oil to alcohol molar ratios [57]. Because all the biodiesel production parameters are closely associated, and the selection of an individual parameter without taking in consideration the rest of the parameters can affect the production process.

TEMPERATURE AND REACTION TIME

Reaction temperature and reaction time depend greatly on each other with an inversely proportional relationship. Both parameters also depend on the reaction speed of the selected catalyst; acid catalysts require higher temperatures and longer reaction times to carry out the conversion. The selection of an optimum temperature range is very important when producing biodiesel, especially when using conventional technologies. Transesterification can occur at different temperatures depending on the oil, without exceeding the boiling point of the alcohol used [58]. Several studies on the kinetic [59-66] of biodiesel production have been performed to select an optimum temperature; however, there are several factors affecting the temperature in the reaction that a specific value has not been delimited.

Maddikeri et al. (2013) [67] investigated the effect of operating temperatures (30 to 50°C) on biodiesel yield using a non-conventional technology, and they noticed that the biodiesel yield increased from 78 to 88% with an increase in temperature from 30 to 40°C; the biodiesel yield only increased by 5% (93%) when the temperature was increased from 40°C to 50°C. It was observed that maximum yield (90%) of biodiesel from waste cooking oil using sonochemical reactors was observed at a molar ratio of 1:12, catalyst concentration of 1.0% and temperature of 40°C [67]. Also, using a non-conventional technology, Hingu et al. (2010) concluded that the optimum conditions for the transesterification process were at a molar ratio of alcohol to oil as 6:1, catalyst concentration of 1 wt %, temperature as 45°C and ultrasound power as 200W with an irradiation time of 40 min [68]. In contrast, Alves et al. (2013) used a conventional heating technology requiring higher temperatures and longer reaction time. The reaction runs were carried out for 2 hours, using oil to alcohol molar ratio of 40:1, temperature range of 60–200°C, catalyst ratio of 1–10% wt., under 700 rpm stirring [56].

Higher temperatures are usually required when using algal biomass as feedstock and conventional heating technologies. Algal oil extraction, as previously discussed, is an energy-intensive process that often demands high temperatures. However, with the extractive-transesterification combination process proposed in this project, extreme temperatures may no longer be necessary. Tsigie et al. (2012) transesterified algal biomass, the optimum results were at a ratio of wet biomass to methanol of 1/4 (g/mL), reaction temperature of 175°C, and a reaction time of 4 h, the reaction product contained 89.71% FAMEs [69]. However, according to El Sherbiny et al. (2010), the reaction temperature should be fixed at 65° or slightly above the boiling point of the used alcohol, in this case methanol [70]. Cheng et al. (2013) [71] performed an extended study on the influence of temperature on biodiesel yield.

They noticed that at room temperature (23°C) the biodiesel yield was 1.23% and the biodiesel yield significantly improved when the reaction temperature was 60°C; however, when the temperature exceeded 60°C, the biodiesel yield did not significantly improved [71].

Conventional Heating Technologies for Biodiesel Production

Conventional heating technologies for biodiesel production using WCO are less complex than that of using algal biomass, yet energy-intensive. Therefore, the following section will emphasize mainly on the conventional heating technologies for algal biodiesel production; however, several of these technologies could be also used when using WCO. WCO conventional heating technologies mainly consist of heating plates, hot water bath, or room temperature reactions. Transesterification of oil feedstock can be performed in the laboratory by conventional heating and mixing methods such as jacketed reactors, oil and sand baths mixing provided by mechanical stirring [72].

Non-Conventional Heating Technologies for Biodiesel Production

Non-conventional technologies for biodiesel production include microwave and ultrasound irradiations, non-catalytic and solvent free reactions, and sub and supercritical methods. These methods proved to be advantageous over conventional methods of jacketed heating, water and oil bath heating. The vast majority of these methods reduce energy consumption, reaction time, and chemical and volume use. However, supercritical transesterification [73] requires higher volume of solvent/reactant, which is difficult to recover. Also, even though supercritical method has the benefit of no catalyst required; it may be disadvantageous due to the adverse process economics as well as safety concerns related to the reaction condition [74]. A non-catalytic biodiesel production route with supercritical methanol allows a simple process and high lipid yield because of simultaneous transesterification of triglycerides and methyl esterification of fatty acids [20, 75]. A fluid is considered supercritical when its temperature and pressure go above its critical point [21]. Another supercritical method is the supercritical carbon dioxide ($SCCO_2$) extraction, which has low toxicity of the supercritical fluid, tuneable solvating power, favorable mass transfer equilibrium due to intermediate diffusion/viscosity properties of the fluid, and the production of solvent-free extract [76-80]. In addition to conventional methods, pyrolysis decomposes biomass under the condition oxygen deficiency and high temperatures. This process was first used for the production of bio-oils or bio-gases from lignocelluloses. Fast pyrolysis is a new technology, which produces bio-fuel in the absence of air at atmospheric pressure with a relatively low temperature (450-550°C) and short heating rate (10^3-10^4°Cs^{-1}) as well as short gas residence time to crack into short chain molecules and be cooled to liquid rapidly.

Transesterification through non-conventional technologies is usually referred to *in-situ* transesterification. The *in-situ* transesterification method has the potential to simplify the conversion process of disrupting the rigid cell wall, in the case of algal biomass, reducing the number of unit operations and consequently the overall process costs and consequently the overall process costs and final biodiesel product costs [81-83]. In situ transesterification is an efficient way to convert oil-bearing biomass to biodiesel directly; hence, elimination the extraction step which is required in the conventional method [84]. During this process, the alcohol may weaken the cellular and lipid body membranes to facilitate the FAME conversion [82]. Interestingly, in situ transesterification do not degrade the protein in the feedstock meal, while they may degrade components found on the meal. Additionally, in situ transesterification reduces the amount of phospholipids to a level below detection [85-86].

Microwave and ultrasound irradiations are the most promising heating technologies to convert lipids into biodiesel. They are both non-conventional heating techniques that have demonstrated that they can tremendously enhance WCO biodiesel production. Microwaves and ultrasound can benefit from shorter extraction and transesterification times and low solvent requirements [87].

MICROWAVE AND ULTRASOUND IRRADIATIONS FOR BIODIESEL PRODUCTION

It is well accepted by now that microwaves and ultrasound produce superior results in many chemical syntheses attributed to both thermal and specific non-thermal effects induced by the irradiation. More specifically, microwaves provide rapid and convenient heating for chemical synthesis but have mass transfer limitations whereas the ultrasound produces intense physical mixing by cavitation but lacks the ability to generate/induce high thermal energy for chemical synthesis. These heat and mass transfer limitations and/or requirements can be addressed collectively by combining the two different techniques in a single reactor. Microwaves require mechanical mixing to enhance mass/heat transfer and to reduce localized superheating whereas an increase in the reaction temperature enhances ultrasound mediated reactions.

While the two technologies were utilized individually to improve the biodiesel synthesis process in numerous studies thus far, the combined effect of these two non-conventional methods has not been studied in detail to date [5, 67-68, 88-93]. Chemat et al. (2006) demonstrated the first study of esterification of propanol using microwave and ultrasound reactor and pyrolysis of urea [94]. In that study, they reported an increase in the yields for both the reactions. Cravotto et al. reported the combined use of the two mechanisms on transesterification of vegetable oil and algal oil extraction [95]. They reported that the ultrasound and microwave either alone or combined have greatly improved the extraction yields along with higher yields and shorter reaction times. However, the combined effect of these two technologies on transesterification reaction has not been reported so far. In addition, none of the studies have attempted to compare the synergistic effect of these non-conventional technologies in terms of their performance. Synergism can be defined as a phenomenon resulting from the effect of a combination of technologies, tools, or reagents that exceeds the sum of their individual effects [95-97].

As previously mentioned, the most commonly used homogeneous catalysts are alkaline metal hydroxides such as sodium hydroxide and potassium hydroxide (NaOH and KOH result in soap formation, which makes the process inefficient by reducing alkyl ester yield and interrupting glycerol recovery [98]. Although, there are many advantages with heterogeneously catalyzed methanolysis (transesterification) reaction, it is very complex because it occurs in a three-phase system consisting of a solid (heterogeneous catalyst) and two immiscible liquid phases (oil and methanol). But, by applying microwaves and ultrasound simultaneously, the heterogeneity of the reaction mixture (or diffusion limitations) can be greatly reduced. Ultrasound provides for intense mixing that is more effective than conventional mechanical mixing. The Microwave effect will be greatly enhanced when the mass transfer limitation is eliminated by simultaneous ultrasound mixing providing uniform and rapid heating throughout the reaction mixture. In the past, to overcome this mass transfer problem in heterogeneous catalysts, co-solvents were introduced to promote miscibility of oil and methanol [99]. Tetrahydrofuran (THF), dimethyl sulfoxide (DMSO), n-hexane, and ethanol were the more commonly used co-solvents in transesterification of vegetable oils with methanol and solid catalysts [100]. Another way to mitigate mass transfer problems associated with heterogeneous catalysts is using structure promoters or catalyst supports which provide more specific surface area and pores for active species where they can anchor and react with large triglyceride molecules [101]. However, in simultaneous microwave and ultrasound mediated reactions, such addition of co-solvents or surface modifications are not required.

Microwave Irradiation

Microwaves are high frequency radio waves that can cause certain molecules to vibrate around a billion times per second [102]. Microwave irradiation is an unconventional energy source, and it has been proven that transesterification with a base homogeneous catalyst is significantly accelerated under microwave irradiation [81]. This advantageous technology has been reported to be the most simple and most effective method for WCO conversion [42, 97]. Using microwave irradiation as a selective and instantaneous heating method has many obvious advantages over conventional technologies for preparing biodiesel. Microwave irradiation is also a faster and more efficient heating process, which directly contributes to molecular diffusion and mass transfer. In addition, methanol is strong microwave absorption solvent. Thus, microwave irradiation can promote the transesterification reaction.

Microwave heating is a non-contact heat source, which heats the whole sample volume simultaneously as compared to conductive heating after being reported for extraction of chemicals from environmental matrices. Microwaves transfer the energy into the materials by dipolar polarization, ionic conduction, and interfacial polarization mechanisms and cause localized rapid and superheating of the reaction materials. If a molecule possesses a dipole moment, then, when it is exposed to microwave irradiation, the dipole tries to align with the applied electric field [103]. Since the electric field is oscillating, the dipoles constantly try to realign to follow this movement. At 2.45 GHz, molecules have time to align with the electric field but not to follow the oscillating field exactly. This continual reorientation of the molecules results in friction and consequently releasing thermal energy. If a molecule is charged, then the electric field component of the microwave irradiation moves the ions back and forth through the sample while also colliding them into each other. This movement again generates heat. In addition, since the energy is interacting with the molecules at a higher rate at which the molecules do not have the time to relax, the heat generated at a microscopic level by this principle can be, for short times, much greater than the overall recorded temperature of the bulk reaction mixture. The microwave interaction with the reaction compounds (triglycerides and alcohol) results in large reduction of activation energy due to increased dipolar polarization phenomenon [104]. This is achieved due to molecular level interaction of the microwaves in the reaction mixture resulting in dipolar rotation and ionic conduction [105-106]. The amount, by which the activation energy is reduced, is essentially dependent on the medium and reaction mechanism [104, 107].

Ultrasound Irradiation

Ultrasound can be a valuable tool for the transesterification of fatty acids, aiming to prepare the biodiesel fuel at industrial scale [108]. The association of the cold methods with ultrasound extraction has contributed to the increase the amount of oil extracted from the microalgal biomass. The chemical effect of cavitation is the generation of highly reactive radicals due to dissociation of the entrapped vapor molecules in the cavitation bubble. These form bubbles, filled with solvent and solute vapor and dissolved gases, which grow and recompress. Under proper conditions, ultrasonic cavitation leads to implosive cavitation bubble collapse, producing intense local heating, high pressures and very short lifetimes. The collapse of the cavitation bubbles gives rise to acoustic microstreaming or formation of small eddies that increase the mass and heat transfer in the liquid, and cause velocity gradients that result in shear stresses [109]. Furthermore, the implosion of the cavitation bubbles caused by ultrasound inside and outside an organism may contribute to rupture of cell walls [110-111] when using algal biomass as feedstock. In addition, the formation, growth, and implosive collapse of micro bubbles induced acoustically in the bulk of the liquid phase increase the mass transfer between the two phases by supplying both heating and mixing [112].

Cavitation causes a localized increase in temperature at the phase boundary and provides the mechanical energy for mixing and the required activation energy for initiating the alcoholysis reaction. The collapse of the cavitation bubbles disrupts the phase boundary and causes emulsification by ultrasonic jets that impinge one liquid to another. These effects speed up the alcoholysis reaction rate and shorten its duration, while high final yields of biodiesel are usually achieved [42, 113-114]. The normal chain alcohols react quite rapidly under ultrasonic irradiation. The velocity of an ultrasonic wave through a material depends on its physical properties. Hence, the ultrasonic velocity decreases with increasing density. Thus, the use of ultrasound enhances the reaction rate and also shifts the equilibrium which results in higher product yields in shorter reaction times. The ultrasound wave frequency seems to affect the reaction rate and the biodiesel yield. The use of ultrasound has an extra advantage as it requires one-third to a half of the energy that is consumed by mechanical agitation [112, 115-116]. Transesterification through ultrasonication can be carried out in a water bath (indirect sonication) or using a horn (direct sonication). Direct sonication has been proven to be more efficient resulting in high biodiesel yield [42, 113-114]. Lin et al. [117] attributed three important phenomena to the effect of ultrasonic irradiation: (1) rapid movement of fluids caused by a variation of sonic pressure causes solvent compression and rarefaction; (2) cavitation; and (3) microstreaming where large amount of vibrational energy is confined in small volumes of reaction medium with less heating [117-118]. Cavitation phenomenon has been found as effective tool for intensification of the esterification of fatty acids [119]. When using ultrasound irradiation, high reaction medium temperatures are not required to carry out the transesterification reaction. In addition, lower methanol to oil molar ratios and smaller amounts of catalyst are required. Ultrasonication enhances the mass transfer between methanol and oil which results in a higher biodiesel yield. It enhances mass transfer and reaction rates in both multiphase reactors and homogeneous systems [120-121].

CONTINUOUS VS. PULSE SONICATION

Many studies reported on the effects of pulse sonication on the transesterification of fresh and used oils in both direct and indirect (via water bath) applications [118, 122-124]. Indirect application was shown to have much lower efficiency reflected by prolonged reaction times when compared to direct ultrasound application [114]. A few studies reported on the direct and continuous applications of ultrasonic processing of biodiesel [68, 114, 123, 125]. The continuous sonication delivers the acoustic radiation continuously such that the time for relaxation among fluid layers in the reaction medium is not possible, thereby increasing thermal energy of the reaction medium. This phenomenon can be considered as loss of some applied energy which could otherwise have been converted to vibrational or cavitational energy. The continuous sonication may energize the transesterification reaction by its thermal effect and is suitable for those reactions requiring thermal excitation. Too much of an exposure to direct sonication may result in undesired reactions and byproduct formation due to over excitation of the reactants and emulsification of reaction compounds often resulting in difficult product separation [114]. On the other hand, the pulse sonication allows for relaxation intervals providing excitation gap to the reaction mixture to form good emulsions without increasing thermal energy of the reaction medium significantly. The pulse sonication is preferred for temperature sensitive chemical reactions. However, the pulse sonication may impact a lower excitation compared to continuous sonication often requiring longer sonication times. In both cases, the direct application of sonication in the reaction mixture has significant effect when compared to indirect sonication applied through water baths and other media. Direct sonic application reduces the reaction time and therefore the energy consumption is reduced significantly as well. Some reactions can be very sensitive to direct sonication and may advance in undesirable direction which needs to be evaluated case by case. For transesterification, the direct sonication may prove to be beneficial with proper reactor design, reaction scheme and optimization process.

Ultrasound cavitation phenomenon can increase the mass transfer between the oil phase and the reactant during the transesterification process [126]. In general, ultrasound replaces the conventional agitation and heating required to establish close contact between the two immiscible phases involved in the reaction, the triglycerides and the reactant (alcohol). According to Qian et al. [127] the wave frequency and specific ultrasonic energy influence biodiesel yields either under pulse or continuous sonication mode within the range of 20 and 48 kHz. The lower is the frequency; the higher will be the biodiesel yield [128]. A wide variety of ultrasonic devices are available to convert triglycerides into fatty acids or biodiesel. These devices may provide direct (applicator or horn) or indirect sonication (i.e., ultrasonic bath or tube type ultrasonicator). Direct sonication is the most effective method, and it has been proven to be more energy-efficient than mechanical mixing [44, 113, 129]. Ultrasound (US) is very effective at dispersing a material present in a solution and its application contributes to a more homogenous reaction mixture. The chemical effects of ultrasound originate from several acoustic phenomena where cavitation (formation, growth, and implosive collapse of bubbles in a liquid) is the most important [130]. The cavities or microbubbles are formed when ultrasound passes through the liquid which consist of both expansion (negative pressure) and compression (positive pressure) [131]. The collapse of cavitation bubbles causes acoustic microstreaming or formation of small eddies that increase the mass and heat transfer in the liquid and causes velocity gradients that result in shear stress [109]. The implosive bubble collapse produces intense local heating, high pressures, and very short lifetimes (less than a nanosecond) that releases a large amount of energy [132]. In ultrasound processing, the temperature of the bulk reaction does not necessarily represent the local microscopic temperature which leads to completion of the transesterification reaction [114, 133]. Hot spots formed in the cavitating bubbles have equivalent temperatures of roughly 5000 K, pressures of about 1000 atm, and very high heating/cooling rates [114, 133]. Thus, ultrasound can create extraordinary reaction environment in otherwise low temperature reaction mixtures [134-135].

Table 3. Physical and chemical properties of vegetable oil methyl esters [4]

Feedstock	V[a]	FP[b]	D[c]	HHV[d]	IV[e]	CN[f]	AV[g]	SN[h]
Soybean	4.08	441	138.7	138.7	138.7	52	0.15	201
Rapeseed	4.3–5.83	453	0.88–0.888	41.55–	–	49–50	0.25–0.45	–
Sunflower	4.9	439	0.88	41.33	142.7	49	0.24	200
Palm	4.42	434	0.86–0.9	41.24	60.07	62	0.08	207
Peanut	4.42	443	0.883	41.32	67.45	54	–	200
Corn	3.39	427	0.88–0.89	41.14	120.3	58–59	–	202
Camelina	6.12-7	–	0.882–0.888	–	152–157	–	0.08–0.52	–
Canola	3.53	–	0.88–0.9	–	103.8	56	–	182
Pumpkin	4.41	–	0.8837	–	115	–	0.48	202
Cotton	4.07	455	0.875	41.18	104.7	54	0.16	204
Jatropha curcas	4.78	–	0.8636	–	108.4	61–63	0.496	202
Pongamina pinnata	4.8	–	0.883	–	–	60–61	0.62	–
Palanga	3.99	–	0.869	–	–	–	–	–
Tallow	–	–	0.856	–	126	59	0.65	244.5
Nile tilapia	–	–	–	–	88.1	51	1.4	–
Poultry	–	–	0.867	–	130	61	0.25	251.23
WCO	4–5.18	148	0.878–0.887	39.26–39.48	–	48	0.15	–

[a] V = Viscosity (at 40°C).
[b] FP = Flash Point (°C).
[c] D = Density (g/cm^3).
[d] HHV = High Heat Value (MJ/Kg).
[e] IV = Iodine Value.
[f] CN = Cetane Number.
[g] AV = Acid Value (mgKOH/g).
[h] SN = Saponification Number.

Table 4. Economics of biodiesel production by supercritical transesterification in US [17]

Plant capacity (tones/year)	125,000	80,000	8,000
Fixed capital	10,395,058	7,953,072	1,997,721
Working capital	1,661,348	1,513,014	313,729
Startup cost	4,984,045	4,539,042	941,187
Total capital cost	17,040,452	14,005,128	3,252,638
Annual variable cost			
Raw material			
WCO	26,068,993	16,784,050	1,709,129
Methanol	4,218,750	2,736,000	278,400
Total raw materials cost	30,287,743	19,520,050	1,987,529
Start up			
Methanol	14,400	9,050	924
Propane	4,409	2,672	269
Total startup cost	18,809	11,722	1,193
Utilities			
Electricity	713,592	456,699	45,670
Cooling water	102,708	67,103	8,217
Biodiesel for the reboiler	978,359	872,157	198,371
Total utilities cost	1,794,660	1,395,958	252,257
By-product credit			
Glycerol	15,973,500	6,234,000	1,017,600
Fixed cost			
Operating labor	1,020,000	1,020,000	1,020,000
Maintenance	485,103	371,143	371,143
Plant overhead	913,021	890,229	890,229
Taxes and insurance	207,901	159,061	39,954
Total fixed cost	2,626,024	2,440,433	2,321,326
Total operating cost	18,789,736	17,134,165	3,538,992
Capital charges[a]	5,314,381	4,353,483	1,207,569
S, G & A	1,205,206	1,074,382	237,328
Required Selling Price	25,309,323	22,562,030	4,983,890
RSP (US$/tones)	202	282	623
RSP (US$/kg)	0.2	0.28	0.26
RSP (US$/l)	0.17	0.24	0.52

[a] 20% return of investment (ROI).

BIODIESEL PROPERTIES

Biodiesel properties vary from feedstock to feedstock. Table 3 shows the physical and chemical properties of vegetable oil methyl ester [4]. Vegetable oils have higher pour and cloud point compared to diesel fuels. Therefore, it is not advisable to use them in winter. However, the cetane number of vegetable oils, in general, is very high hence reducing the ignition delay. In addition, they have a high iodine value that increases its oxidation rate. Therefore, long time storage is not recommended for these types of fuels. The properties also change with the use of alcohol. Ethyl esters are more commonly used a lubricant agents.

Biodiesel Production Costs

A conceptual design of a biodiesel production process using waste cooking oil as feedstock via supercritical transesterification with methanol to methyl esters (biodiesel) was reported [17]. Because waste cooking oil contains water and high free fatty acids, supercritical transesterification was chosen as an advantageous method to eliminate the pre-treatment capital and operating costs. The economics of this process were studied at different production capacities, i.e., 125,000; 80,000 and 8000 tons biodiesel/year (see Table 4). This process is promising because high purity biodiesel can be produced with almost pure glycerol as a byproduct. The economic assessment showed that biodiesel can be sold at US$ 0.17/l (125,000 tons/year), US$ 0.24/l (80,000 tons/year) and US$ 0.52/l for the smallest capacity (8000 tons/year). As expected, the sensitive factors influencing the economic feasibility were raw material price, plant capacity, glycerol price and capital cost.

Conclusion

Biodiesel production can help reduce our dependency on fossil fuel imports, improve environmental sustainability, and economic security at national and global levels. It is crucial to develop energy-efficient and resource-efficient technologies for biodiesel production to be competitive. This means that feedstock derived from waste sources and other marginal oil products such as microbial oils should be considered to improve the process economics. Further reductions in chemical and energy usage can be achieved by developing process intensification techniques such as microwave and ultrasound. All these factors will contribute to the development of process technologies for sustainable biodiesel production.

Acknowledgments

This research was supported by the Office of Research and Economic Development (ORED), the Bagley College of Engineering (BCoE), and the Department of Civil and Environmental Engineering (CEE) at Mississippi State University.

References

[1] Vicente G, Martinez M, Aracil J (2004) Integrated biodiesel production: a comparison of different homogeneous catalysts systems. *Bioresource Technology* 92, 297-305.

[2] Feng P, Yang K, Xu Z, Wang Z, Fan L, Qin L, Zhu S, Shang C, Chai P, Yuan Z, Hu L (2014) Growth and lipid accumulation characteristic of Scenedesmus obliquus in semi-continuous cultivation outdoors for biodiesel feedstock production. *Bioresource Technology* 173, 406-414.

[3] Gerpen VJ, Shanks B, Pruszko R, Clements D, Knothe G. Biodiesel Analytical method. Subcontractor report. *National Renewable Energy Laboratory;* 2004.

[4] Talebian-Kiakalaieh, A., Amin, N. A. S., & Mazaheri, H. (2013). A review on novel processes of biodiesel production from waste cooking oil. *Applied Energy,* 104, 683-710.

[5] Patil PD, Gude VG, Reddy HK, Muppaneni T, Deng S. Biodiesel production from waste cooking oil using sulfuric acid and microwave irradiation processes. *J Environ Prot* 2012; 3:107–13.

[6] Refaat AA (2010). Different techniques for the production of biodiesel from waste vegetable oil. *Int. J. Environ. Sci. Tech,* 7, 183-213.

[7] Demirbas A (2011). Waste management, waste resource facilities and waste conversion processes. *Energy Conversion and Management,* 52, 1280-1287.

[8] Cvengros J, Cvengrosova Z (2004). Used frying oils and fats and their utilization in the production of methyl esters of higher fatty acids. *Biomass Bioenergy,* 27, 173-181.

[9] Kulkarni MG, Dalai AK (2006). Waste cooking oils an economical source for biodiesel: a review. *Industrial & Engineering Chemistry Research,* 45, 2901-2913.

[10] Gnanaprakasam A, Sivakumar VM, Surendhar A, Thirumarimurugan M (2013). Recent strategy of biodiesel production from waste cooking oil and process influencing parameters: a review. *Journal of Energy,* 2013, 1-10.

[11] Phan AN, Phan TM (2008). Biodiesel production from waste cooking oils. *Fuel* 87, 3490-3496.

[12] Tomasevic AV, Marinkovic SS (2003). Methanolysis of used frying oil. *Fuel Processing Technol.* 81, 1-6.

[13] Issariyakul T, Dalai AK (2014) Biodiesel from vegetable oils. *Renewable and Sustainable Energy Reviews,* 31, 446-471.

[14] Singh A, He B, Thompson J, van Gerpen J (2006). Process optimization of biodiesel production using different alkaline catalyst. *Appl. Eng. Agr.,* 22, 597-600.

[15] Freedman B, Butterfield RO, Pryde EH (1986) Transesterification kinetics of soybean oil. *J. Am. Oil Chem Soc.* 63, 1375-1380.

[16] Zhang Y. Dube MA, McLean DD, Kates M (2003a). Biodiesel production from waste cooking oil: 1. Process design and technological assessment. *Bioresource Tech.*, 89, 1-16.

[17] Kasteren JMN, Nisworo AP (2007). A process model to estimate the cost of industrial scale biodiesel production from waste cooking oil by supercritical transesterification. *Resources, Conservation and Recycling,* 50, 442-458.

[18] Galadima A, Muraza O (2014). Biodiesel production from algae by using heterogeneous catalysts: A critical review. *Energy* 1-12.

[19] Velasquez-Orta SB, Lee JGM, Harvey A (2012). Alkaline in situ transesterification of Chlorella vulgaris. *Fuel* 94, 544–550.

[20] Demirbas A (2005) Biodiesel production from vegetable oils via catalytic and non-catalytic supercritical methanol transesterification methods. *Prog Energy Comb Sc* 31, 466–87.

[21] Bharathiraja B, Chakravarthy M, Kumar RR, Yuvaraj D, Jayamuthunagai J, Kumar RP, Palani S (2014) Biodiesel production using chemical and biological methods – A review of process, catalyst, acyl acceptor, source and process variables. *Renewable and Sustainable Energy Reviews* 38, 368–382.

[22] Laurens LML, Quinn M, Van Wychen S, Templeton DW, Wolfrum EJ (2012) Accurate and reliable quantification of total microalgal fuel potential as fatty acid methyl esters by in situ transesterification. *Analytical and Bioanalytical Chemistry* 403, 167-178.

[23] Di Serio M, Cozzolino M, Giordano M, Tesser R, Patrono P, Santacesaria E (2007). From homogeneous to heterogeneous catalysts in biodiesel production. *Ind Eng Chem Res*, 46, 6379–84.

[24] Liu X, He H, Wang Y, Zhu S, Piao X (2008). Transesterification of soybean oil to biodiesel using CaO as a solid base catalyst. *Fuel*, 87, 216–21.

[25] Xie WL, Peng H, Chen LG (2006). Transesterification of soybean oil catalyzed by potassium loaded on alumina as a solid-base catalyst. *Appl Catal A,* 300, 67–74.

[26] Suppes GJ, Dasari MA, Doskocil EJ (2004). Transesterification of soybean oil with zeolite and metal catalysts. *Appl Catal A,* 257, 213–23.

[27] Tsuji H, Yagi F, Hattori H, Kita H (1994). Self-condensation of n-butyraldehyde over solid base catalysts. *J Catalyst,* 148, 759–70.

[28] Cantrell DG, Gillie LJ, Lee AF, Wilson K (2005). Structure-reactivity correlations in MgAl hydrotalcite catalysts for biodiesel synthesis. *Appl Catal A,* 287, 183–90.

[29] Lee DW, Park YM, Lee KY (2009). Heterogeneous base catalysts for transesterification in biodiesel synthesis. *Catal Surv Asia,* 13, 63–77.

[30] Di Serio M, Tesser R, Pengmei L, Santacesaria E (2008). Heterogeneous catalysts for biodiesel production. *Energy Fuels,* 22, 207–17.

[31] Schuchardt U, Sercheli R, Varga RM (1998). Transesterification of vegetable oils: a review. *J Braz Chem Soc,* 9, 199–210.

[32] Smith VG, Notheisz F. Heterogeneous catalysis in organic chemistry. New York, NY, USA: Academic Press Inc.; 2006.

[33] Bender ML. Mechanisms of catalysis of nucleophilic reactions of carboxylic acid derivatives. *Chem Rev* 1960; 60:53–113.

[34] Demirbas A (2008). Comparison of transesterification methods for production of biodiesel from vegetable oils and fats. *Energy Convers Manage,* 49, 125–30.

[35] Yan S, Di Maggio C, Mohan S, Kim M, Salley SO, Simon KY (2010). Advancements in heterogeneous catalysis for biodiesel synthesis. *Top Catal;* 53:721–36.

[36] Chorkendorff I, Niemantsverdriet JW. Concepts of modern catalysis and kinetics. Weinheim, Germany: Wiley-VCH; 2003.

[37] Refaat AA (2011). Biodiesel production using solid metal oxide catalysts. *Int J Environ Sci Technol,* 8, 203–21.

[38] Freedman B, Pyrde EH, Mounts TL (1984). Variables affecting the yields of fatty esters from transesterified vegetable oils. *J Am Oil Chem Soc,* 61, 1638–43.

[39] Bradley DC, Mehrotra RC, Gaur DP. Metal alkoxides. New York: Academic Press; 1978.

[40] Martinez-Guerra E, Gude VG (2014) Transesterification of used vegetable oil catalyzed by barium oxide under simultaneous microwave and ultrasound irradiations. *Energy Conversion and Management* 88, 633-640.

[41] Felizardo P, Neiva Correia MJ, Raposo I, Mendes JF, Berkemeier R, Moura Bordado J (2006). Production of biodiesel from waste frying oils. *Waste Manage.,* 26, 487–494.

[42] Andrade Torres E, Cerqueira GS, Ferrer TM, Quintella CM, Raboni M, Torretta V, Urbini G (2013). Recovery of different waste vegetable oils for biodiesel production: a pilot experience in Bahia State, Brazil. *Waste Manage.* 33, 2670–2674.

[43] Encinar JM, Gonzalez JF, Rodrıguez JJ, Tejedor A (2002). Biodiesel fuels from vegetable oils: transesterification of Cynara cardunculus L. Oils with ethanol. *Energy Fuels,* 16, 443–450.

[44] Martinez-Guerra E, Gude VG (2014) Transesterification of waste vegetable oil under pulse sonication using ethanol, methanol and ethanol-methanol mixtures. *Waste Management* 34, 2611-2620.

[45] Encinar JM, González JF, Rodríguez-Reinares, A (2007). Ethanolysis of used frying oil. Biodiesel preparation and characterization. *Fuel Process.* Technol., 88, 513–522.

[46] Likozar B, Levec J (2014a). Transesterification of canola, palm, peanut, soybean and sunflower oil with methanol, ethanol, isopropanol, butanol and tert-butanol to biodiesel: modelling of chemical equilibrium, reaction kinetics and mass transfer based on fatty acid composition. *Appl. Energy,* 123, 108–120.

[47] Narvaez PC, Sanchez FJ, Godoy-Silva RD (2009). Continuous methanolysis of palm oil using a liquid–liquid film reactor. *J. Am. Oil Chem. Soc.,* 86, 343–352.

[48] Stamenkovic OS, Todorovic´ ZB, Lazic´ ML, Veljkovic´ VB, Skala DU (2008). Kinetics of sunflower oil methanolysis at low temperatures. *Bioresour. Technol.,* 99 (5), 1131–1140.

[49] Kai T, Mataki Y, Nakazato T, Takanashi H (2010). Optimization of reaction conditions of two-step batch operation for biodiesel fuel production using KOH catalyst. *J. Chem. Eng. Jpn*., 43 (1), 90–94.

[50] Leung DY, Wu X., Leung MKH (2010). A review on biodiesel production using catalyzed transesterification. *Appl. Energy*, 87 (4), 1083–1095.

[51] Likozar B, Levec, J (2014b). Effect of process conditions on equilibrium, reaction kinetics and mass transfer for triglyceride transesterification to biodiesel: experimental and modeling based on fatty acid composition. *Fuel Process. Technol.* 122, 30–41.

[52] Pukale DD, Maddikeri GL, Gogate PR, Pandit AB, Pratap AP (2015). Ultrasound assisted transesterification of waste cooking oil using heterogeneous solid catalyst. *Ultrasonics Sonochemistry*, 22, 278-286.

[53] Leung DYC, Guo Y (2006). Transesterification of neat and used frying oil: Optimization for biodiesel production. *Fuel Processing Technol.* 87, 883-890.

[54] Meng X, Chen G, Yonghong W (2008). Biodiesel production from waste cooking oil vial alkali catalyst and its engine test. *Fuel Processing Technol.,* 89, 851-857.

[55] Dehkordi AM, Ghasemi M (2012) Transesterification of waste cooking oil to biodiesel using Ca and Zr mixed oxides as heterogeneous base catalysts. *Fuel Processing Technol.*, 97, 45-51.

[56] Alves, CT, Andrade HMC, Santos RCD, de Melo SAB, Torres EA (2013). Transesterification reaction using zinc aluminates as heterogeneous catalyst and supercritical carbon dioxide. *Chemical Engineering Transactions,* 32, 883-888.

[57] Zheng S, Kates M, Dube MA, McLean (2006). Acid-catalyzed production of biodiesel from waste frying oil. *Biomass and Bionergy,* 30, 267-272.

[58] Ferrari RA, Pighinelli ALMT, Park KJ (2005). Chapter 10: Biodiesel production and quality in Biofuel's Engineering Process Technology. *Open access* 241-240.

[59] Jain S, Sharma MP, Rajvanshi S (2011). Acid-base catalyzed transesterification kinetics of waste cooking oil. *Fuel Processing Technol.,* 92, 32-38.

[60] Neumann K, Werth K, Martin A, Gorak A (2015). Biodiesel production from waste cooking oils through esterification: Catalyst screening, chemical equilibrium and

reaction kinetics. *Chemical Engineering Research and Design,* doi:10.1016/j.cherd.2015.11.008.
[61] Adewale P, Dumont MJ, Ngadi M (2015). Enzyme-catalyzed synthesis and kinetics of ultrasonic-assisted biodiesel production from waste tallow. *Ultrasonics Sonochemistry*, 27, 1-9.
[62] Chang J, Zhang LB, Yang CJ, Ye W, C hen J, Peng JH, Srinivasakannan C, Li W (2015). Kinetics of microwave roasting of zinc slag oxidation dust with concentrated sulfuric acid and water leaching. *Chemical Engineering and Processing*, 97, 75-83.
[63] Michelin S, Penha FM, Sychoski MM, Scherer RP, Treichel H, Valerico A, Di Luccio M, de Oliveira D, Oliveria JV (2015). Kinetics of ultrasound-assisted enzymatic biodiesel production from Macauba coconut oil. *Renewable Energy*, 76, 388-393.
[64] Patil PD, Reddy H, Muppaneni T, Schaub T, Holguin FO, Cooke P, Lammers P, Nirmalakhandan N, Li Y, Lu X, Deng S (2013). In situ ethyl ester production from wet algal biomass under microwave-mediated supercritical ethanol conditions. *Bioresource Technol.,* 139, 308-315.
[65] Lin R, Zhu Y, Tavlarides LL (2013). Mechanism and kinetics of thermal decomposition of biodiesel fuel. *Fuel,* 106, 593-604.
[66] Kostic MD, Bazargan A, Stanenkovic OS, Veljkovic VB, McKay G (2016). Optimization and kinetics of sunflower oil methanolysis catalyzed by calcium oxide-based catalyst derived from palm kernel shell biochar. *Fuel,* 163, 304-313.
[67] Maddikeri GL, Pandit AB, Gogate PR. Ultrasound assisted interesterification of waste cooking oil and methyl acetate for biodiesel and triacetin production. *Fuel Process Technol* 2013; 116:241–9.
[68] Hingu SM, Gogate PR, Rathod VK. Synthesis of biodiesel from waste cooking oil using sonochemical reactors. *Ultrason Sonochem* 2010; 17:827–32.
[69] Tsigie YA, Huynh HL, Ismadji S, Engida AM, Ju YH (2012) In situ biodiesel production from wet Chlorella vulgaris under subcritical condition. *Chemical Engineering Journal* 213, 104–108.
[70] El Sherbiny SA, Refaat AA, Sheltawy ST (2010). Production of biodiesel using the microwave technique. *Journal of advanced Research*, 1, 309-314.
[71] Cheng J, Yu T, Li T, Zhou J, Cen K (2013) Using wet microalgae for direct biodiesel production via microwave irradiation. *Bioresource Technology* 131, 531–535.
[72] Grant GE, Gude VG (2014). Kinetics of ultrasonic transesterification of waste cooking oils. *Environ Prog Sustain Energy*, 33, 1051–8.
[73] Nan Y, Liu J, Lin R, Tavlarides LL (2015) Production of biodiesel from microalgae oil (Chlorella protothecoides) by non-catalytic transesterification in supercritical methanoland ethanol: Process optimization. *J. of Supercritical Fluids* 97, 174–182.
[74] Takisawa K, Kanemoto K, Miyazaki T, Kitamura Y (2013) Hydrolysis for direct esterification of lipids from wet microalgae. *Bioresource Technology,* 144, 38–43.
[75] Demirbas A (2002) Diesel fuel from vegetable oil via transesterification and soap pyrolysis. *Energy Sources,* 24, 835–41.
[76] Macias-Sanchez MD, Mantell C, Rodriguez M, de la Ossa EM, Lubian LM, Montero O (2007) Supercritical fluid extraction of carotenoids and chlorophyll a from Synechococcus sp. *J of Supercritical Fluids*, 39, 323–329.

[77] Mendes RL, Reis AD, Palabra AF (2006) Supercritical CO2 extraction of gamma-linolenic acid and other lipids from Arthrospira (Spirulina) maxima: Comparison with organic solvent extraction. *Food Chemistry,* 99, 57–63.
[78] Pourmortazavi SM, Hajimirsadeghi SS (2007) Supercritical fluid extraction in plant essential and volatile oil analysis – review. *J of Chromatography A,* 1163, 2–24.
[79] Taylor LT (1996) Supercritical Fluid Extraction. John Wiley & Sons Inc, New York.
[80] Thana P, Machmudah S, Goto M, Sasaki M, Pavasant P, Shotipruk A (2008) Response surface methodology to supercritical carbon dioxide extraction of astaxanthin from Haematococcus pluvialis. *Bioresource Technology,* 99, 3110–3115.
[81] Martinez-Guerra E, Gude VG, Mondala A, Holmes W, Hernandez R (2014) Extractive-transesterification of algal lipids under microwave irradiation with hexane as solvent. *Bioresource Technology,* 156, 240-247.
[82] Hass MJ, Wagner K (2011) Simplifying biodiesel production: The direct or in situ transesterification algal biomass. *European Journal of Lipid Science and Technology,* 11, 1219-1229.
[83] Jazzar S, Olivares-Carrillo P, Pérez de los Ríos A, Marzouki MN, Acién-Fernández FG, Fernández-Sevilla JM, Molina-Grima E, Smaali I, Quesada-Medina J (2015) Direct supercritical methanolysis of wet and dry unwashed marine microalgae (Nannochloropsis gaditana) to biodiesel. *Applied Energy,* 148, 210–219.
[84] Kim GV, Choi WY, Kang DH, Lee SY, Lee HY (2014) Enhancement of biodiesel production from marine alga, Scenedesmus sp. through in situ transesterification process associated with acidic catalyst. *Biomed Res Int.,* 2014, 1-11.
[85] Wahlen BD, Willis RM, Seefeldt LC (2011) Biodiesel production by simultaneous extraction and conversion of total lipids from microalgae, cyanobacteria, and wild mixed-cultures. *Bioresource Technology*, 102, 2724–2730.
[86] Vicente G, Bautista LF, Rodriguez R, Gutierrez FJ, Sadaba I, Ruiz-Vazquez RM, Torres-Martinez S, Garre V (2009) Biodiesel production from biomass of an oleaginous fungus. *Biochemical Engineering Journal* 48, 22-27.
[87] Gude VG, Grant GE, Patil P, Deng S (2013) Biodiesel from low cost renewable feedstock. *Open Engineering* 3, 595-605.
[88] Babajide O, Petrik L, Amigun B, Ameer F (2010). Low-cost feedstock conversion to biodiesel via ultrasound technology. *Energies,* 3, 1691–703.
[89] Kumar D, Kumar G, Poonam CP, Singh (2010). Ultrasonic-assisted transesterification of Jatropha curcas oil using solid catalyst, Na/SiO2. *Ultrason Sonochem,* 17, 839–44.
[90] Teixeira LSG, Assis JCR, Mendonc DR, Santos ITV, Guimaraes PRB, Pontes LAM (2009). Comparison between conventional and ultrasonic preparation of beef tallow biodiesel. *Fuel Process Technol,* 90, 1164–6.
[91] Refaat AA, El Sheltawy ST (2008). Comparing three options for biodiesel production from waste vegetable oil. *WIT Trans Ecol Environ* 2008; 109:133–40.
[92] Patil PD, Gude VG, Camacho LM, Deng S (2010). Microwave-assisted catalytic transesterification of Camelina sativa oil. *Energy Fuels*, 24, 1298–304.
[93] Gude VG, Patil PD, Deng S, Nirmalakhandan N. Microwave enhanced methods for biodiesel production and other environmental applications. In: Sanghi R, Singh V, editors. Green chemistry for environmental remediation. New York: *Wiley Interscience*; 2011. p. 209–49.

[94] Chemat F, Poux M, Di Martino JL, Berlan J (2006). An original microwave-ultrasound combined reactor suitable for organic synthesis: application to pyrolysis and esterification. *J Microw Power Electromagn Energy,* 31, 19–22.

[95] Cravotto G, Boffa L, Mantegna S, Perego P, Avogadro M, Cintas P (2008). Improved extraction of vegetable oils under high-intensity ultrasound and/or microwaves. *Ultrason Sonochem,* 15, 898–902.

[96] Cravotto G, Cintas P (2007). The combined use of microwaves and ultrasound: improved tools in process chemistry and organic synthesis. *Chem Eur J,* 13, 1902–9.

[97] Martinez-Guerra E, Gude VG (2014) Synergistic effect of simultaneous microwave and ultrasound irradiations on transesterification of waste vegetable oil. *Fuel* 137, 100-108.

[98] Freedman B, Butterfield RO, Pryde EH (1986). Transesterification kinetics of soybean oil. *J Am Oil Chem Soc*, 63, 1375–80.

[99] Alhassan Y, Kumar N, Bugaje IM, Pali HS, Kathkar P (2014). Co-solvents transesterification of cotton seed oil into biodiesel: effects of reaction conditions on quality of fatty acids methyl esters. *Energy Convers Manage,* 84, 640–8.

[100] Gryglewicz S (1999). Rapeseed oil methyl esters preparation using heterogeneous catalysts. *Bioresour Technol,* 70, 249–53.

[101] Zabeti M, Daud WMAW, Aroua MK. Activity of solid catalysts for biodiesel production: a review. *Fuel Process Technol* 2009; 90:770–7.

[102] Cui Y, Liang Y (2014) Direct transesterification of wet Cryptococcus curvatus cells to biodiesel through use of microwave irradiation. *Applied Energy* 119, 438–444.

[103] Gude PD, Patil PD, Deng S, Nirmalakhandan N (2011) Microwave enhanced methods for biodiesel production and other environmental applications. *Green Chemistry for Environ. Remediation*, Wiley Publishers, 209-249.

[104] L. Perreux A, Loupy (2001) A tentative rationalization of microwave effects in organic synthesis according to the reaction medium, and mechanistic considerations. *Tetrahedron,* 57, 9199–9223.

[105] Lidstrom P, Tierney JP, Wathey B, Westman J (2001) Microwave assisted organic synthesis: A review Tetrahedron 57, 9225–9283.

[106] Kanitkar AV (2010) Parameterization of microwave assisted oil extraction and its transesterification to biodiesel. Master's Thesis. Lousiana State University, Baton Rougue, LA.

[107] Patil PD, Gude VG, Mannarswamy A, Cooke P, Nirmalakhandan N, Lammers P, Deng S (2012) Comparison of direct transesterification of algal biomass under supercritical methanol and microwave irradiation conditions. *Fuel* 97, 822-831.

[108] Stavarache C, Vinatoru M, Nishimura R, Maeda Y (2005) Fatty acids methyl esters from vegetable oil by means of ultrasonic energy. *Ultrasonics Sonochemistry* 12 (2005) 367–372.

[109] Adewuyi YG (2001) Sonochemistry: environmental science and engineering applications. *Ind Eng Chem Res* 40, 4681–715.

[110] Mason TJ, Paniwnyk L, Lorimer JP (1996). The use of ultrasound in food technology. *Ultrason. Sonochem.* 3, S253–S256.

[111] Araujo GS, Matos LJBL, Fernandes JO, Cartaxo SJM, Gonçalves LRB et al. (2013) Extraction of lipids from microalgae by ultrasound application: Prospection of the optimal extraction method. *Ultrasonics Sonochemistry* 20, 95–98.

[112] Avramović JM, Stamenković OS, Todorović ZB, Lazić ML, Veljković VB (2010) The optimization of the ultrasound-assisted base-catalyzed sunflower oil methanolysis by a full factorial design. *Fuel Processing Technology* 91, 1551-1557.

[113] Martinez-Guerra E, Gude VG (2015) Continuous and pulse sonication effects on transesterification of used vegetable oil. *Energy Conversion and Management* 96, 268-276.

[114] Grant GE, Gude VG (2013) Biodiesel from waste cooking oils via direct sonication. *Applied Energy* 109, 135-144.

[115] Ji J, Wang J, Li Y, Yu Y, Xu Z (2006) Preparation of biodiesel with the help of ultrasonic and hydrodynamic cavitation, *Ultrasonics,* 44, 411–414.

[116] Lifka J, Ondruschka B (2004) Influence of mass transfer on the production of biodiesel, *Chem. Eng. Technol.* 27, 1156–1159.

[117] Lin JR, Yen TF (1993). An upgrading process through cavitations and surfactance. *Energy Fuels,* 7, 111–8.

[118] Kumar D, Kumar G, Poonam, Singh CP (2010). Ultrasonic-assisted transesterification of Jatropha curcus oil using solid catalyst, Na/SiO2. *Ultrason Sonochem,* 17, 839–44.

[119] Parkar PA, Choudhary HA, Moholkar VS (2012). Mechanistic and kinetic investigations in ultrasound assisted acid catalyzed biodiesel synthesis. *Chem Eng J,* 187, 248-60.

[120] Bar R (1988). Ultrasound-enhanced bioprocesses: cholesterol oxidation by Rhodococcus erythropolis. *Biotechnol Bioeng,* 32, 655–63.

[121] Chisti Y (2003). Sonobioreactors: using ultrasound for enhanced microbial productivity. *Trends Biotechnol*, 21, 89–93.

[122] Thanh LT, Okitsu K, Sadanaga Y, Takenaka N (2010). Ultrasound-assisted production of biodiesel fuel from vegetable oils in a small scale circulation process. *Bioresour Technol*, 101, 639–45.

[123] Choudhury HA, Goswami PP, Malani RS, Moholkar VS (2014). Ultrasonic biodiesel synthesis from crude Jatropha curcas oil with heterogeneous base catalyst: mechanistic insight and statistical optimization. *Ultrason Sonochem*, 21, 1050–64.

[124] Pal A, Kachhwaha SS (2013). Biodiesel production of waste cooking oil through ultrasound cavitation. *Int J Eng Res Technol*, 6, 291–306.

[125] Gharat N, Rathod VK (2013). Ultrasound assisted enzyme catalyzed transesterification of waste cooking oil with dimethyl carbonate. *Ultrason Sonochem*, 20, 900–5.

[126] Fayyazi E, Ghobadian B, Najafi G, Hosseinzadeh B, Mamat R, Hosseinzadeh J (2015), An ultrasound-assisted system for the optimization of biodiesel production from chicken fat oil using a genetic algorithm and response surface methodology. *Ultrasonics Sonochemistry,* 26, 312-320.

[127] Qian WW, Han PF, Lu XP (2010). KF/CaO as solid base catalyst for transesterification to biodiesel by ultrasound, *Ranlio Huaxue Xuebao,* 38, 52-56.

[128] Rokhina EV, Lens P, Virkutyte J (2009). Low frequency ultrasound in biotechnology: State of the art, *Trends in Biotechnology* 27, 298-306.

[129] Martinez-Guerra E, Gude VG, Mondala A, Holmes W, Hernandez R (2014) Microwave and ultrasound enhanced extractive-transesterification of algal lipids. *Applied Energy* 129, 354-363.

[130] Masjedi-Arani M, Salavati-Niasari M (2016). A simple sonochemical approach for synthesis and characterization of Zn_2SiO_4 nanostructures. *Ultrasonics Sonochemistry,* 29, 226-235.

[131] Lee SB, Lee JD, Hong IK (2011). Ultrasonic energy effect on vegetable oil based biodiesel synthetic process. *J. Ind. Eng. Chem.*, 17, 138-143.

[132] Zinatloo-Ajabshir S, Salavati-Niasari M (2014). Synthesis of pure nanocrystalline ZrO2 via s simple sonochemical-assisted route. *J. Ind. Eng. Chem.* 20, 3313-3319.

[133] Takase M, Chen Y, Liu H, Zhao T, Yang L, Wu S (2014). Biodiesel production from non-edible Sylibum marianum oil using heterogeneous solid base catalyst under ultrasonication. *Ultrasonics Sonochemistry* 21, 1752-1762.

[134] Suslick KS (1989). The chemical effects of ultrasound, *Sci. Am.,* 260, 80-86.

[135] Gude VG (2015). Synergism of microwaves and ultrasound for advanced biorefineries. *Resour.-Efficient Technol.,* 1, 116–125.

In: Vegetable Oil
Editor: Brittany Holt

ISBN: 978-1-63485-128-2

Chapter 4

VEGETABLE OIL AS A FEEDSTOCK FOR BIODIESEL SYNTHESIS

Ivana Lukić[1,*], Željka Kesić[1], Miodrag Zdujić[2] and Dejan Skala[1]

[1]University of Belgrade, Faculty of Technology and Metallurgy, Belgrade, Serbia

[2]Institute of Technical Sciences of the Serbian Academy of Sciences and Arts, Belgrade, Serbia

ABSTRACT

Limited reserves of fossil fuels as well as the growing concern for the environment, has led to a worldwide search for renewable energy sources, among which biodiesel, a mixture of fatty acid methyl esters (FAME), is one of the most perspective alternative fuels since it is a non-toxic and can be produced from different renewable sources through simple cost-effective alcoholysis, while being compatible with existing infrastructures. Vegetable oils, as renewable in nature and environmentally friendly, with a possibility to be produced on a large scale, represent a promising feedstock for biodiesel production. In this chapter, a comprehensive review of different vegetable oils as a feedstock for biodiesel synthesis is reported, including edible and non-edible oils, as well as waste vegetable oils.

Selection of feedstock for biodiesel production mainly depends on the specific conditions and circumstances in some region (climate, presence of certain crops, the economic development of a country, etc.). Various fatty acid compositions of triacylglycerols directly determine the quality and fulfillment of the standards of biodiesel. One of the crucial points which determine technology route for biodiesel synthesis is content of free fatty acids (FFA) which might be present in vegetable oils, as well as the presence of water and other compounds. Also, many analysis performed in the past have shown that the production cost of biodiesel is mainly determined by the price of used feedstock, which represents 70–80% of total production costs of biodiesel.

* Corresponding author. Tel.: +381 11 3303 709; E-mail address: ivanal888@yahoo.co.uk.

Currently more than 95% of feedstock comes from edible oils since they are easily accessible, consists mainly of triacylglycerols, whereby the properties of biodiesel produced from these oils are suitable to be used as diesel fuel substitute. Most commonly used edible oils for biodiesel production are rapeseed, soybean, sunflower and palm. However, for economic and social reasons, in recent years research and development of biodiesel production has focused on other sources of triacylglycerols, in order to replace edible oils by lower-cost non-edible plant oils and the waste cooking oils, feedstocks that are unsuitable for human consumption.

Properties of different oils and biodiesel obtained from them as well as the technologies suitable for biodiesel synthesis are compared in this chapter. The well-known fact is that the conventional and to-day widely applied homogeneous method of biodiesel synthesis is sensitive to the presence of impurities in the oil, primarily the presence of FFA and water. Furthermore, biodiesel synthesis is followed by creation of large amount of wastewater produced during neutralization of catalyst and purification of final product. The drawbacks of a homogeneous process can be avoided by applying technologies based on utilization of heterogeneous catalyst or by application of the non-catalytic supercritical process of biodiesel synthesis. These technologies for biodiesel production were also analyzed and compared.

Keywords: biodiesel, transesterification, vegetable oil, catalysis

INTRODUCTION

Fossil fuels, primarily crude oil and its derivatives, represent the most important source of energy for transport because of the large heat capacity, widespread crude oil resources and relatively simple exploitation and manipulation. However, there is a great concern over the depletion of the earth's reserves of crude oil and a great impact this would have on a societies dependent on such type of energy. Moreover, a huge disadvantage of crude oil derivatives consumption is also associated with the environmental pollution caused by the emission of huge amounts of carbon dioxide, nitrogen and sulfur oxides, as well as soot and other solid particles produced in the combustion of these fuels.

Consequently, in the past few decades, great attention has been focused on finding opportunities to reduce consumption of fossil fuels and replace them with alternative and renewable energy sources. Alternative fuel must be technically appropriate, economically competitive, based on renewable raw materials, environmentally beneficial and easily accessible (Srivastava and Prasad, 2000). From the technical aspect, biodiesel represents a high quality fuel for diesel engines. It is quite similar to diesel fuel in its main characteristics, and since being compatible, biodiesel and diesel fuel can be blended in any proportion and used in unmodified or only slightly modified engines. Biodiesel contains 10 to 11% oxygen by weight, which may encourage better combustion in an engine than diesel fuels (Srivastava and Prasad, 2000). However, the presence of oxygen in fatty acid methyl esters, compounds which are the main constituents of biodiesel, leads to a lower heating value of biodiesel (around 12%) compared to diesel fuel, which directly affects the higher consumption of fuel. Relatively high flash point, of about 150°C (while that of diesel is 70°C), is advantage of biodiesel, because it is non-flammable, non-explosive and makes handling, transport and storage easier (Demirbaş, 2009a). Biodiesel provides better ignition and has better lubricant properties. From the standpoint of environmental protection, biodiesel has many advantages.

First of all, it is non-toxic and biodegradable; it degrades about four times faster than diesel fuel (Demirbaş, 2007a). After 28 days all the biodiesels were 77–89% biodegraded, while diesel fuel was only 18% biodegraded (Demirbaş, 2007a). Using biodiesel in diesel engines significantly reduce the emissions of particulate matter, carbon monoxide and unburned hydrocarbons, due to higher cetane number and higher oxygen contents compared with diesel fuel (Mofijur et al., 2013). Combustion of biodiesel also provides reduction in CO_2, SO_2, aromatic and polyaromatic compounds formation. Only NOx emission is increased when biodiesel is used as fuel in the compression ignition engine. Very important fact regarding the environmental impact of biodiesel is that the emissions of CO_2 generated during combustion in engines are used in process of vegetable photosynthesis. The CO_2 is released into the atmosphere when the biodiesel is burned and it is recycled by the growing plants, which are later processed into the fuel.

The idea of using vegetable oil as fuel for diesel engines arose more than hundred years ago, in the year 1900, when Rudolf Diesel demonstrated the operation of diesel engines powered by peanut oil at the Paris World Exposition (Knothe et al., 2005). Although demonstration was successful, the project has not been implemented and the interest in plant oils diminished, since it was a time of the fast development of crude oil production. The use of vegetable oils as diesel engine fuel is documented in literature from the 1920s through the 1940s, most often with the theme of providing tropical colonies of European countries with an independent source of fuel (Knothe, 2010). Palm oil was probably the feedstock that received the most attention in mentioned period of time, although cottonseed and some other oils were also tested (Knothe et al., 2005). Besides that, vegetable oils were used as fuels only in emergency conditions, such as during the World War II (Knothe, 2001).

In spite of satisfactory performance of vegetable oils as fuel for diesel engine, later studies showed that they create some engine problems especially in direct-injection engines. The major drawback of vegetable oils is their high viscosity that is about an order of magnitude greater than that of conventional diesel fuel (Knothe 2001). A high viscosity leads to poor atomization of the fuel in the engine's combustion chambers, incomplete combustion, coking of the fuel injectors, ring carbonization, and ultimately results in operational problems, leading to increase of the carbonaceous solid deposit in engine (Ma and Hanna, 1999; Schwab et al., 1987). These problems can be solved by either adapting the engine to the fuel, which is more complicated and expensive, or by adapting the fuel to the engine. The latter led to modification of vegetable oils by various technologies to produce adequate fuel for existing diesel engines (Schwab et al., 1987). The four possible modification procedures have been investigated: alcoholysis or transesterification, pyrolysis, dilution with conventional diesel fuel and microemulsification (Schwab et al., 1987). The most common and, according to the results, the best way of vegetable oils modification is alcoholysis, the reaction between vegetable oil triacylglycerols (TAG) and lower alcohol, usually methanol or ethanol, which leads to fatty acid methyl or ethyl esters formation, called simplified "biodiesel" when used as fuel. Probably the first documented use of a fuel that corresponds to the present definition of biodiesel is registered in a Belgian patent from 1937 that describes the use of palm oil ethyl esters (Chavanne, 1937). A year later a bus running between Brussels and Louvain was operated on biodiesel (Issariyakul and Dalai, 2014). The energy crises of the 1970s and early 1980s and growing ecological awareness led to the renewal of interest in vegetable oil derived fuels (Knothe, 2010). The first patent applications on the use of vegetable oil methyl esters as diesel fuel substitutes were reported in 1980 (Hartman, 1980; Tanaka et al, 1980). The first research on the possibilities of production and use of biodiesel was reported in 1981 in South Africa and then in 1982 in Austria, Germany and New Zealand (Körbitz, 1999). Already in 1985 a small pilot plant in Austria tested the production of rapeseed oil methyl esters and the year 1996 was a big step forward in biodiesel production with the start-up of large industrial scale plants in France and Germany (Körbitz, 1999).

FEEDSTOCK FOR BIODIESEL SYNTHESIS

Vegetable oils represent the main lipid feedstock, i.e., source of TAG for biodiesel production. They are renewable in nature and environmentally friendly, with a possibility to be produced on a large scale. Different vegetable oils, including edible and non-edible oils, as well as waste vegetable oils have been tested as a feedstock for biodiesel production. Other potential materials that could be used for biodiesel synthesis are animal fat and more recently oil obtained from microalgae. The main disadvantage of animal fat, such as by-products from meat and fish industry, pork and poultry fat, beef tallow and fish oil (Atabani et al., 2012; Balat et al., 2011), is a high content of saturated fatty acids, which affect the quality of their methyl esters at low temperatures. Although microalgae, considered to be the third generation feedstock, have shown potential for high biomass growth and lipid productivity and could be grown on non-arable land, there are some constraints that have to be overcome in order to make it economical and sustainable source for biodiesel production. Due to the composition of their oil which widely varies between species and the fact that microalgae in addition to TAG may also contain substances which cannot be converted to methyl esters, the yield and the quality of produced biodiesel is not satisfactory; energy-consuming, and therefore costly extraction of oil from algae is also another disadvantage of their usage as a feedstock (Singh et al., 2014).

Vegetable oil can be obtained from more than 350 different oil-bearing crops (Demirbaş, 2009a). Oils are extracted mainly from seeds and fruits of plants and consist of mixtures of organic compounds that contain, depending on their origin, approximately 98% of TAG and small amounts of monoacylglycerols and diacylglycerols, in addition to free fatty acids, phospholipids, carotenes, tocopherols, water and other impurities. Generally, factors such as geography, climate, soil conditions and economics of specific country determine which vegetable oil is of most interest for potential use for biodiesel production (Knothe, 2001). In Europe, rapeseed (canola) oil is the most commonly used as a feedstock; in the United States it is soybean oil and in Southeast Asia palm oil (Demirbaş, 2009a). The oil content and the yield per hectare from the crops are important factors that determine the suitability of a feedstock for biodiesel production, since oil crops with higher yield are more preferable because they can reduce the production cost (Atabani et al., 2012; Gui et al., 2008).

Another important criterion that should be considered when choosing the feedstock for biodiesel production is the composition of the oil. The composition of oil will subsequently determine the properties and the quality of obtained biodiesel, but also will affect the choice of the production technology. The major components of vegetable oils are TAGs, esters of fatty acids and glycerol. Fatty acids vary in carbon chain length and in the number of unsaturated (double) bonds. The percentage and type of fatty acid composition depends mainly on the plant species and their growth conditions, and the fatty acids that are the most often present in TAG of both non-edible and edible oils are oleic (C 18:1), linoleic (C 18:2), stearic (C 18:0) and palmitic acid (C 16:0) (Atabani et al., 2013).

Also, very important fact is that the price of the lipid feedstock is the major factor determining biodiesel price (Grabovski and McCormick, 1998). The price of feedstock constitutes approximately 70–85% of the total biodiesel production cost when even the least expensive refined vegetable oils are used (Demirbaş, 2007a; Gui et al., 2008, Knothe et al., 2005).

Edible Oils

Currently, more than 95% of biodiesel produced globally is from edible vegetable oils, such as rapeseed, soybean, sunflower and palm (Gui et al., 2008). Edible oils are easily accessible because of the abundant agricultural production, and their major components are TAG while impurities are reduced to a minimum. Therefore, edible oils represent the most suitable raw material for biodiesel production, taking into account their availability, ease of processing and the quality of obtained biodiesel. Edible oils are considered as the first generation biodiesel feedstock. The oil content and yield, fatty acid composition and other properties of edible oils are given in Table 1.

Soybean Oil

Soybean (*Glycine max*) is an annual crop from the *Leguminosae* family. It is one of the world's most important sources of plant protein and edible oil for both humans and animals (Milazzo et al., 2013b). The major producers of soybean oil are the United States, Brazil and Argentina and this oil is now the world's second largest oilseed in terms of total oil production, being on the first place until recently (http://apps.fas.usda.gov/psdonline/circulars/oilseeds.pdf). With high protein content (about 40%), which is commonly used for animal feed, soybean is the highest yielding source of vegetable protein (up to 50%) globally (Milazzo et al., 2013b). The oil content in soybean seed ranges from 15% to 22% (Issariyakul and Dalai, 2014), which is one of the drawbacks of soybean oil as biodiesel feedstock. Another disadvantage is that soybean is a low-yielding oilseed crop with an oil yield from 1.8 to 3.6 t/ha (Milazzo et al., 2013b).

The major fatty acids in soybean oil are oleic (C18:1), linoleic (C18:2) and linolenic (C18:3). Content of polyunsaturated fatty acids is a characteristic that determines oxidative stability of fuel, and with its high content (about 63%) soybean oil obtained biodiesel has the problem of low resistance to oxidation (Milazzo et al., 2013b). This problem can be overcome, since soybean oil is often modified by traditional plant breeding, chemical or genetic manipulation which have been used to modify the fatty acid composition of soybean oil in order to improve its oxidative or functional properties (Wang, 2002). Soybean biodiesel offers enhanced biodegradation, reduced toxicity, increased flash point and lubricity and lower emissions (Milazzo et al., 2013b).

Rapeseed and Canola Oil

Rapeseed is a bright-yellow flowering member of the family *Brassicaceae* (mustard or cabbage family), first cultivated in India almost 4000 years ago. Large-scale planting was first reported in Europe in 13th century (Przybylski and Mag, 2002). According to the United States Department of Agriculture, rapeseed was the third-leading source of vegetable oil in the World in 2015, after palm and soybean oil; it was the world's second-leading source of protein meal (http://apps.fas.usda.gov/psdonline/circulars/oilseeds.pdf).

Early rapeseed cultivars had high levels of erucic acid (C22:1) in the oil and high levels of glucosinolates in the meal, components that were considered to be a health concern (Przybylski and Mag, 2002). This was the reason for plant breeding programs initiated in Canada, which led to the first low-erucic acid cultivars of *Brassica napus* and *Brassica rapa* (Przybylski and Mag, 2002). By the definition, canola should contain less than 2% erucic acid in the oil and less than 30 μmol/g glucosinolates in the air-dried, oil-free meal (Przybylski and Mag, 2002).

Rapeseed oil was used in the very first efforts to promote biodiesel as a fuel obtained from renewable sources. Rapeseed and canola seeds have oil content over 40% in which the dominant fatty acids include oleic acid (C18:1), linoleic acid (C18:2) and erucic acid (C22:1) (Issariyakul and Dalai, 2014). The type of fatty acid in the feedstock oil determines the biodiesel characteristics, and a higher percentage of unsaturated fatty acids will result in biodiesel having better cold flow properties (Gui et al., 2008). Due to its high content of oleic acid (about 64%) biodiesel produced from rapeseed and canola oil has very good cold flow properties (Gui et al., 2008), making this biodiesel suitable fuel for usage during the winter. Good fuel performance and more oil per unit of land area compared to other oil sources make rapeseed oil the preferred oil feedstock for biodiesel production in most of Europe. However, according to some studies, rapeseed methyl ester is currently not compatible with diesel oil particulate filters and may also not be considered as an optimal prospective fuel in view of its boiling characteristic which is less beneficial for fuels ignition and its combustion in the cylinder (Millazo et al., 2013a). Also, there is a great concern for the use of rapeseed oil for biodiesel production because rapeseed is presently grown with a high level of nitrogen containing fertilizer, thus biodiesel made of these oils generates N_2O, one of undesired compounds which causes greenhouse effect (Karmakar et al., 2010).

Sunflower Oil

The sunflower is a plant native to the Americas, where was domesticated long before arrival of European explorers (Gupta, 2002). Cultivated sunflower (*Helianthus annuus*) was introduced into Europe by the Spanish in the sixteenth century (Gupta, 2002). Sunflower oil is the fourth most important vegetable oil in the world and accounts for approximately 10% of the total world's consumption of plant-derived edible oil (Balalić et al., 2012). Oil is derived from the seed of the sunflower plant, which contains 48–52% of high quality edible oil (Balalić et al., 2012).

The fatty acid composition of sunflower oil is strongly influenced by both genetic and geographic location and climatic condition (Lajara et al., 1990). Usually, sunflower oil comprises up to 90% unsaturated fatty acids (combined oleic, C18:1, and linoleic, C18:2) and approximately 10% saturated fatty acids (palmitic, C16:0, and stearic, C18:0) (Zheljazkov et al., 2011). Several varieties of sunflower oilseeds have been developed by standard plant breeding methods, mainly to vary the amount of oleic and linoleic acid. Beside the traditional sunflower oil with linoleic acid content of 65–70% of the oil, there are mid-oleic acid varieties (55–70% oleic acid and 15–35% linoleic acid) and high-oleic acid varieties (>80% oleic acid and only 5–9% linoleic acid) (Gupta, 2002). High content of linoleic acid in traditional sunflower oil has unfavorable effect on the quality of biodiesel, due to less stability to oxidation, but still sunflower oil is the second most common feedstock for biodiesel production in Europe.

Palm Oil

The oil palm (*Elaeis guineensis*) originated from South Africa and was introduced to East Asia as an ornamental plant in 1884 (Lin, 2002). Another species of oil palm, *Elaeis oleifera,* originates from Central and South America, but its oil is more unsaturated and this species gives less oil yield, making it uneconomical to plant on a commercial scale (Lin, 2002). The oil palm is the most efficient oil-producing plant, with about 5.4 L of oil per hectare per year (Lin, 2002), which is five to ten times higher than oil yields from soybeans, sunflower or rapeseeds. Palm oil recently overtook soybean oil and now takes the first place in the list of oils produced in the world. Two South East Asian countries dominate production of palm oil: Indonesia with 56% and Malaysia with 32% of all palm oil production (http://apps.fas.usda.gov/psdonline/circulars/oilseeds.pdf).

Two types of oil, with different fatty acid composition and oil yield, are obtained from the oil palm fruit (*Elaeis guineensis*): palm oil from the mesocarp (palm oil in the narrow sense) and palm kernel oil from the kernel inside the nut (Lin, 2002). There are great differences between palm oil and palm kernel oil with respect to their physical and chemical characteristics, especially in their fatty acid content. Palm oil has a balanced fatty acid composition in which the level of saturated fatty acids is almost equal to that of the unsaturated fatty acids (Lin, 2002). Palmitic acid (44–45%) and oleic acid (39–40%) are the major component acids of TAG along with linoleic acid (10–11%) and only a trace amount of linolenic acid. Due to the low content of linoleic and absence of linolenic acid, the methyl esters from palm oil have high oxidation stability (Mekhilef et al., 2011), but on the other hand, they have poor cold flow properties because of the high content of saturated fatty acids (Gui et al, 2008). Palm kernel oil belongs to so called 'lauric' oils and contains mainly lauric acid (C12:0, about 48%), myristic acid (C14:0, about 16%) and oleic acid (about 15%), while no other fatty acids are present at more than 10% (Pantzaris and Basiron, 2002).

Other Edible Oils

Beside these four edible oils that are most commonly used, some other oils such as corn, cottonseed, coconut, peanut, linseed, sesame, almond etc. could be used as a feedstock for biodiesel synthesis. However, all listed oils have a high production and trade price, thus it is unlikely that they will find industrial application in the production of biodiesel.

Table 1. Oil content and yield, the fatty acid composition and other properties of edible oils

Oil	Oil content, %	Oil yield, L/ha/year	Fatty acid, %						Acid value, mg KOH/g	Iodine value, gI_2/100g oil	Saponification value, mg KOH/g	Reference
			16:0	18:0	18:1	18:2	18:3	other				
Soybean	15–20	446	13.9	2.1	23.2	56.2	4.3	–	0.2	128-143	195.3	Atabani et al. (2012); Demirbaş (2003); Karmakar et al. (2010)
Rapeseed	38–46	1190	4.6	1.6	33.0	20.4	7.4	23[a]	1.14	108.1	197.1	Atabani et al. (2012); Demirbaş (2003)
Canola	–	–	4.3	1.7	61.0	20.8	9.3	1.2	<0.5	–	189.8	Issariyakul and Dalai, (2014); Karmakar et al. (2010)
Sunflower	25–35	952	6.4	2.9	17.7	72.9	–	–	0.15	132.3	191.7	Atabani et al. (2012); Demirbaş (2003)
Palm	30–60	5950	42.6	4.4	40.5	10.1	0.2	1.1	1.4	48-58	208.6	Atabani et al. (2012); Demirbaş (2003); Karmakar et al. (2010)
Corn	48	172	11.8	2.0	24.8	61.3	–	0.3	0.11	119.4	194.1	Demirbaş (2003); Ma and Hanna (1999)
Cottonseed	18–25	325	28.7	0.9	13.0	57.4	–	–	0.07	113.2	207.7	Atabani et al. (2012); Demirbaş (2003); Ma and Hanna (1999)
Olive	45–70	1212	5.0	1.6	74.7	17.6	–	0.8	<2	100.2	196.8	Atabani et al. (2012); Demirbaş (2003)
Peanut	45–55	1059	11.4	2.4	48.3	32	0.9	4.0[b]	0.2	119.6	199.8	Atabani et al. (2012); Demirbaş (2003); Ma and Hanna (1999)
Linseed	40–44	–	5.1	2.5	18.9	18.1	55.1	–	8.3	156.7	188.7	Atabani et al. (2012); Aransiola et al. (2014); Demirbaş (2003)
Sesame	–	696	13.1	3.9	52.8	30.2	–	–	2.4–10.2	91.8	210.3	Atabani et al. (2012); Aransiola et al. (2014); Demirbaş (2003)
Coconut	63–65	2689	8.9	2.7	6.4	1.6	–	65.9[c]	11.6	7.5–10.5	267.6	Demirbaş (2003); Karmakar et al. (2010); Pantzaris and Basiron (2002)

[a]Erucic acid.
[b]Peanut oil contains about 2.7% of 22:0 and 1.3% of 24:0 fatty acids.
[c]47.8% of 12:0 and 18.1% of 4:0 fatty acids.

Cotton (*Gossypium hirsutum* L.) is an important crop that yields the natural fiber used by the textile industry. Cottonseed oil is a relatively low cost feedstock for biodiesel production as cotton seeds are considered as by-products or waste (Karmakar et al., 2010). The fatty acid profile of cottonseed oil is typical of the oleic–linoleic group of vegetable oils, since these two unsaturated fatty acids make up almost 75% of the total fatty acids. Oleic makes up about 22%, palmitic around 24%, linoleic about 52% and linolenic acid content is usually less than 1% (O'Brien, 2002).

The coconut palm is the species *Cocos nucifera* and the coconut oil is derived from copra, which is the dried kernel or 'meat'of coconuts (Pantzaris and Basiron, 2002). The largest producing countries are the Philippines and Indonesia, and coconut oil production is about 3% of total oils. Coconut oil belongs to so called 'lauric' oils, together with palm kernel oil, because lauric acid (C12:0) is the major fatty acid in this oil. These oils are characterized by high level of the shorter and medium fatty acid chain lengths (C6–C14), of about 80%, while in the non-lauric vegetable oils they are below 2%. The major fatty acids are lauric (about 48%) and myristic (about 18%), while no other fatty acids are present at more than about 8% (Pantzaris and Basiron, 2002). Coconut oil remains solid at relatively higher temperature than most of other vegetable oils. The main drawback of using coconut oil in engines is that it starts solidifying at a temperature below 22°C whereby at 14°C it does not flow at all (Karmakar et al., 2010).

Peanut oil (*Arachis hypogeae*) is native to South America, Mexico and Central America. The physicochemical characteristics of peanut oil biodiesel closely resemble to those of diesel fuel. Peanut oil produces approximately 1170 L biodiesel/ha, compared to 475 L for soybean oil (Karmakar et al., 2010). However, the production of biodiesel from peanut oil is not economically viable as peanut oil is more valuable than soy oil in the world's market.

Corn (*Zea mays*) is a crop mainly sown in USA for its starch and protein content. It is not practically viable to grow this crop specifically for biodiesel production as the extraction process cannot produce a grade of oil which is suitable enough for production of biodiesel (Karmakar et al., 2010). But when the crop is turned into ethanol after fermentation, the oil can easily be separated and viably used for biodiesel production.

Linseed (*Linum usitatissimum*) originated from Mediterranean coastal countries and is cultivated in Canada, Argentina, India and USA. Linseeds are a source of high quality proteins, soluble fiber and a high content of polyunsaturated fatty acids. Main use of linseed oil is as industrial oil based on its high unsaturation, but increasingly it is consumed as food oil. The biodiesel produced from linseed oil shows high relative density because of presence of more than 25% of unsaturated esters with more than two double bonds (Karmakar et al., 2010).

Non-Edible Oils

Although today biodiesel is mainly produced from edible oils which are easily available on large scale from the agricultural industry, there are numerous reasons against their use as a feedstock for biodiesel production. First of all, this refers to the food versus fuel debate, i.e., sociological question of whether to use oil as food or for fuel production. Namely, edible oils are a source of important nutrients in the human diet. There are concerns that large-scale production of biodiesel from edible oils may bring global imbalance to the food supply and demand market. The problem is also that biodiesel is competing limited fertile land availability with food industry for plantation of oil crop – arable land that would otherwise have been used to grow food would instead be used to grow fuel. Also, the use of food raw materials for biofuels increases their cost, and the increase in food prices mostly affects the poorest populations of a society. But the increase in raw material prices has also a negative impact on the sustainability of biodiesel production as well. Another issue that should not be ignored is that the extensive production of oil crops that could be used for the production of biofuels leads to the devastation of the soil and deforestation in many countries in order to increase the available fertile land for plantation purposes.

Therefore, non-edible vegetable oils or the second-generation feedstock have become more attractive as a promising, cheaper, raw material for sustainable biodiesel production. There are a number of plants that produce significant amounts of non-edible oils which are not suitable for human consumption due to the presence of toxic compounds and could grow in the barren lands. However, the main disadvantage of use of non-edible oils for biodiesel production is that most of them contain a high content of free fatty acids, which increases the biodiesel production cost (Banković-Ilić et al., 2012). Some of these non-edible oilseed crops include jatropha tree (*Jatropha curcas*), karanja (*Pongamia pinnata*), castor (*Ricinus communis*), tobacco seed (*Nicotiana tabacum*), mahua (*Madhuca indica*), neem (*Azadirachta indica*), rubber seed (*Hevea brasiliensis*), sea mango (*Cerbera odollam*), etc. The oil content and yield, fatty acid composition and other properties of non-dible oils are given in Table 2.

Jatropha Curcas Oil

Jatropha tree (*Jatropha curcas*) is a tropical plant belonging to the *Euphorbiaceae* family (Atabani et al., 2012). The presence of some toxic components renders this oil unsuitable for use in cooking, but makes it very attractive for fuel production. It can grow almost anywhere and under a wide variety of climatic conditions. Jatropha is grown in marginal and waste lands, on sandy, saline and degraded soils, without the need to compete with the land for food production. It is a drought-resistant plant and due to its characteristics can be easily cultivated without intensive care and with minimal efforts (Gui et al., 2008).

Recently Jatropha is being considered as one of the most promising potential oil sources for biodiesel production in South-East Asia, Central and South America, India and Africa (Banković-Ilić et al., 2012). Jatropha oil content varies depending on the types of species, climatic conditions and mainly on the altitude where it is grown (Karmakar et al., 2010), but seeds contain about 40–60% and kernels 46–58% of oil (Banković-Ilić et al., 2012; Kumar and Sharma, 2011). Jatropha produces oil rich in oleic (42%) and linoleic (35%) acid and smaller amounts of palmitic (14%) and stearic acid (6%), but fatty acid composition could be altered to some extent through interspecific hybridization (Kumar and Sharma, 2011).

Karanja (*Pongamia pinnata*) Oil

Karanja (*Pongamia pinnata*) is a fast growing tree belonging to the *Legumnosae* family, with the potential for high oil seed production and the benefit of the ability to grow on marginal land. These properties support the suitability of this plant for large-scale vegetable oil production required by a sustainable biodiesel industry (Balat et al., 2011). *Pongamia Pinnata* is native in Indian subcontinent and South-East Asia and has been successfully introduced to humid tropical regions of the world as well as parts of Australia, New Zealand, China and the USA (Atabani et al., 2013). It is one of the few nitrogen-fixing trees and can be cultivated on land that has been exhausted of nutrients, helping to improve the soil quality so that the exhausted land can be reused for agricultural purpose (Gui et al., 2008). *Pongamia* tree produces seeds with significant oil content, but the oil has many toxic substances that do not allow its use as cooking oil. The seed oil content ranges between 30 and 40% (Atabani et al., 2013) and the predominant fatty acid in oil is oleic acid (51.8%), followed by linoleic acid (17.7%), palmitic acid (10.2%), stearic acid (7.0%), and linolenic acid (3.6%) (Balat et al., 2011). In addition to these fatty acids, *Pongamia* oil also contains eicosenoic acid in reasonable amounts (9.5–12.4%).

Castor (*Ricinus communis*) Oil

Castor plant (*Ricinus communis*) belongs to the *Eurphorbiaceae* family and is non-edible oilseed crop that is easily grown and resistant to drought. The tree is native to India, China, Brazil, some countries of former Soviet Union and Thailand. India produces about 60% of the world castor production (Banković-Ilić et al., 2012). Castor seed is high in oil and contains about 46–55% (Atabani et al., 2013). Castor oil has the most unique composition with approximately 89.5% ricinoleic acid which is also known as castor oil acid. Ricinoleic acid is an unsaturated fatty acid which is soluble in most organic solvents (Gui et al., 2008). Due to its composition, castor oil is the only oil completely soluble in alcohol and does not require heat and the consequent energy requirement of other vegetable oils in transforming them into fuel (Kumar and Sharma, 2011). Viscosity of castor oil is up to 7-time higher than that of other vegetable oils (Banković-Ilić et al., 2012), but despite it, the kinematic viscosity of transesterified castor oil is comparable to other vegetable oils making it suitable as a biodiesel blend (Kumar and Sharma, 2011).

Tobacco Seed (*Nicotiana tabacum*) Oil

Tobacco (*Nicotiana tabacum* L.) seed oil is a by-product of tobacco leaf production that contains significant amount of oil (35–49% by weight). It can be cultivated in more than 100 countries worldwide such as Macedonia, Turkey, South Serbia and widespread in North and South America (Atabani et al., 2013). The oil is non-edible, but physical, chemical and thermal properties are comparable with other vegetable oils, therefore it has the potentiality to be considered as a new feedstock for biodiesel production.

Mahua (*Madhuca indica*) Oil

Mahua (*Madhuca indica*) is a tropical tree from the *Sapotaceae* family found largely in the central and northern plains and forests of India. The kernel constitutes about 70% of the seed and contains 50% oil with a relatively high percentage of saturated fatty acids such as palmitic (17.8%) and stearic (14.0%) acids (Kumar and Sharma, 2011). The remaining fatty acids are mainly distributed among unsaturated components such as oleic (46.3%) and linoleic (17.9%) acids. The relatively high percentage of saturated fatty acids (35.8%) found in mahua oil results in relatively poor low-temperature properties (Kumar and Sharma, 2011). The mahua oil generally contains about 20% of free fatty acids, therefore converting this oil to biodiesel requires proper procedure and technology (Kumar and Sharma, 2011).

Neem (*Azadirachta indica*) Oil

Neem (*Azadirachta indica*) tree belongs to the *Meliaceae* family, can grow in almost all kinds of soil including clay, saline, alkaline, dry and stony soils, and can tolerate some extreme conditions like temperature of 45°C and rainfall as low as 250 mm (Atabani et al., 2013). Neem tree is native to India, Pakistan, Sri Lanka, Burma, Malaysia, Indonesia, Japan, and the tropical regions of Australia. Almost the whole tree is usable for various purposes such as medicines, pesticides and organic fertilizer (Banković-Ilić et al., 2012). The seed of the fruit contains 20–30 wt% and kernels 25–45% of oil (Atabani et al., 2013), generally with high amounts of free fatty acids, similar to mahua oil, which affect the process of biodiesel production.

Rubber Seed (*Hevea brasiliensis*) Oil

Rubber tree (*Hevea brasiliensistree*) belongs to the family *Euphorbiaceae* and originates from the Amazon rain forest (Brazil). The tree is the primary source of natural rubber and produces 99% of world's natural rubber. The oil content of the seeds ranges from 50 to 60% and of kernel from 40 to 50% of oil which may contain up to 17% FFA (Atabani et al., 2013). The oil is high in unsaturated constituents such as linoleic (39.6%), oleic (24.6%), and linolenic (16.3%) acids, and other fatty acids found in rubber seed oil include saturated species such as palmitic (10.2%) and stearic (8.7%) acids (Kumar and Sharma, 2011).

Sea Mango (*Cerbera odollam*) Oil

Sea mango (*Cerbera odollam*) is a tree belonging to the poisonous *Apocynaceae* family. It is well-known as a 'suicide tree' because of its highly poisonous nature and toxic content in the seed is transferred into the oil after the extraction process. However, the toxin can be easily separated out from the extracted oil by decantation (Kumar and Sharma, 2011). The oil content from *Cerbera odollam* seeds is 54%. The fatty acid composition of *Cerbera odollam* oil is mainly oleic (48.1%), followed by palmitic (30.3%), linoleic (17.8%) and stearic (3.8%) (Atabani et al., 2013), and the free fatty acid content in *Cerbera odollam* oil is significantly higher, as well as in the majority of non-edible oils.

Table 2. Oil content and yield, the fatty acid composition and other properties of non-edible oils

Oil	Oil content, %	Oil yield, L/ha/year	Fatty acid, %						Acid value, mg KOH/g	Iodine value, $gI_2/100g$ oil	Saponification value, mg KOH/g	Reference
			16:0	18:0	18:1	18:2	18:3	other				
Jatropha curcas	Seed: 40–60, kernel: 46–58	1892	12.7–17.0	5.5–9.7	39.1–44.7	32.1–41.6	0.2	–	1.2–45	82–98	193–208	Atabani et al. (2013); Balat (2011); Karmakar et al. (2010)
Karanja (*Pongamia pinnata*)	Seed: 25–50, kernel: 30–50	225–2250[a]	3.7–7.9	2.4–8.9	44.5–71.3	10.8–18.3	–	1.1–3.5	2.53–20	81–90	185–195	Atabani et al. (2013); Karmakar et al. (2010); Srivastava and Prasad (2000)
Castor (*Ricinus communis*)	46–55	1188	1.1	3.1	4.9	1.3	–	89.6[b]	3.0	83–86	191.1	Atabani et al. (2013); Demirbaş (2003); Karmakar et al. (2010)
Tobacco seed (*Nicotiana tabacum*)	Seed: 36–41 kernel: 17	2825	10.9	3.3	14.5	69.5	0.7	0.8	–	125–154	191.5	Atabani et al. (2013); Singh and Singh (2010)
Mahua (*Madhuca indica*)	Seed: 35–50, kernel: 50	–	16.0–28.2	14.0–25.1	41.0–51.0	8.9–17.9	–	0.0–3.3	–	58–70	190.5	Atabani et al. (2013); Srivastava and Prasad (2000)
Neem (*Azadirachta indica*)	Seed: 20–30, kernel: 25–45	2670[a]	13.6–16.2	14.4–24.1	49.1–61.9	2.3–15.8	0.8–3.4	0.2–0.26	32.6	65–80	209.7	Atabani et al. (2013); Aransiola et al. (2014); Srivastava and Prasad (2000)
Rubber seed (*Hevea brasiliensis*)	Seed: 50–60, kernel: 40–50	–	10.2	8.7	24.6	39.6	16.3	–	1.7–42	133.9–142.5	183.9–226	Aransiola (2014); Atabani et al. (2013); Onoji et al. (2016)
Sea mango (*Cerbera odollam*)	Seed: 54, kernel: 6.4	–	24.9–30.3	3.8–5.8	48.1–52.8	13.6–17.8	0.08	2.7	12.8	–	–	Atabani et al. (2013); Kansedo and Lee (2012)

[a]kg/ha/year.
[b]Castor oil contains 89.5% of ricinoleic acid.

Table 3. The fatty acid composition and other properties of waste vegetable oils

Oil	Fatty acid, %							Acid value, mg KOH/g	Iodine value, $gI_2/100g$ oil	Saponification value, mg KOH/g	Reference
	16:0	16:1	18:0	18:1	18:2	18:3	other				
Waste cooking oil from different sources	8.5	–	3.1	21.2	55.2	5.9	4.2	3.6	83	207	Yaakob et al. (2013)
	7.64	0.07	8.68	18.38	62.11	1.7	1.42	–	–	–	Srilatha et al. (2012)
	11.58	0.11	4.26	24.84	53.55	5.6	3.78	7.6	–	–	Yan et al. (2009)
	5.18	0.51	2.1	59.7	19.31	6.82	6.37	–	–	–	Chhetri et al. (2008)
	6.8	–	3.1	33.7	56.4	–	–	2.24	–	204.6	Gupta et al. (2015)
	26.5	–	21.35	10.9	1.7	–	–	–	–	–	Talebian-Kiakalaieh et al. (2013)
	39.3	0.18	2.3	46.3	11.9	–	–	5.0	–	183.5	Lam and Lee (2010)
	20.4	4.6	4.8	52.9	13.5	0.8	2.0	2.1	–	–	Leung andGuo (2006)
	32.4	–	3.8	42.5	15.3	5.9	–	0.98	–	183.4	Dehkordi and Ghasemi (2012)
	29.6	–	3.7	48.6	18.0	trace	–	1.04	–	–	Viola et al. (2012)

Waste Vegetable Oils

In order to avoid all the above-mentioned disadvantages of using edible oils as a feedstock for biodiesel production, and to find cheap and more economical alternative, waste cooking oils attracted great attention of researchers. Waste vegetable oils are generally low in cost, their price is two to three times lower than the price of refined vegetable oils, therefore, the usage of waste vegetable oils can reduce biodiesel production costs by 60–90% (Talebian-Kiakalaieh et al., 2013). The fatty acid composition and other properties of waste vegetable oils from different sources are given in Table 3.

Huge quantities of waste cooking oils are available throughout the world, especially in the developed countries, and estimated potential amount of waste cooking oil collected in the EU is about 0.7–1 million tons/yr (Kulkarni and Dalai, 2006), in the United States about 10 millions tones/yr and in China 4.5 millions tones/yr (Maddikeri et al., 2012). Waste cooking oils are mainly collected from restaurants, hotels, households or other large food processing and service facilities. Most of these oils were used for deep-frying process, after which they are no longer suitable for human consumption and their disposal is followed by special regulation. Previously, waste cooking oil was used as an ingredient in animal feed. However, since 2002, the European Union has banned the use of this oil in fodder making, which is likely to be extended worldwide (Maddikeri et al., 2012). Even though some of waste cooking oil is used for soap production (Chhetri et al., 2008), a major part of it is discharged into the environment, causing contamination of the water and land resources. Using waste vegetable oils to produce biodiesel, beside the fact that they reduce the cost of biodiesel production, also could solve the problem of waste material disposal, thus achieving the reduction of environmental pollution.

However, properties of biodiesel produced from waste cooking oil would be largely dependent on the physicochemical properties of such feedstock. Heat treatment of vegetable oils at high temperatures (160–200°C) causes some of major physical and chemical changes in the oil, which more or less differs from the oil composition depending on temperature and duration of heating. Some common physical changes observed in vegetable oil after its use for frying are an increase in the viscosity and the specific heat, a change in color and the surface tension, and an increase in the tendency to foam. Chemical changes are the consequence of thermolytic, oxidative and hydrolytic reactions that take place during the heating of oil at high temperature (Maddikeri et al., 2012). A thermolytic reaction occurs in the absence of oxygen at very high temperatures where series of normal alkanes, alkenes, lower fatty acids, symmetric ketones, oxopropyl esters, CO, and CO_2 is produced from saturated fatty acids, while unsaturated fatty acids form basically dimeric compounds, including dehydrodimers, saturated dimers, and polycyclic compounds (Maddikeri et al., 2012). Oxidative reaction occurs when oxygen in air comes in contact with the oil, forming hydroperoxides as a primary product, which further form many other compounds such as aldehydes, ketones, hydrocarbons, alcohols, acids and esters (Kulkarni and Dalai, 2006; Maddikeri et al., 2012). In the case of hydrolytic reaction, the steam produced during the processing of food containing water causes the hydrolysis of TAG, resulting in the formation of FFA, glycerol, monoacylglycerols and diacylglycerols (Maddikeri et al., 2012).

As a result of all these reactions, waste vegetable oil contains certain amount of undesirable compounds, primarily solid impurities, FFA and water, whose content may be very different; the free fatty acids varied from 0.7% to 41.8%, and moisture from 0.01% to

55.38% (Çanakçi, 2007). For that reason, the production of biodiesel from waste vegetable oils is not so simple, often requiring pretreatment before subjecting the oil to transesterification. The separation of suspended solids from waste cooking oil can be effectively carried out by filtration or centrifugation and the removal of soluble salts by water washing (Demirbas, 2009b; Maddikeri et al., 2012). Water from the oil can be removed by drying at temperature over 100°C (Demirbas, 2009b; Yaakob et al., 2013) or by adsorption using silica gel, anhydrous sulphates etc. (Maddikeri et al., 2012). For FFA removal several methods have been proposed: a) esterification with methanol in the presence of acid catalyst producing the methyl esters; b) neutralization using alkali (KOH or NaOH) forming soaps that are later decanted; c) adsorption processes as efficient and low energy consuming operation where treatment by activated alumina and silica gel, or magnesium silicate and aluminum oxide were applied for reducing both FFA and water content (Maddikeri et al., 2012). Distillation was also used for drying and deacidification of waste cooking oil, while maintaining the original composition of the feedstock. Pretreatment of waste cooking oil using some of these steps can significantly improve the yield and quality of biodiesel obtained from waste cooking oil. However, the pretreatment step affects the cost of biodiesel production, why all possible methods for biodiesel synthesis from waste vegetable oils should be carefully considered.

TRANSESTERIFICATION (ALCOHOLYSIS) METHODS FOR BIODIESEL SYNTHESIS

Transesterification (also called alcoholysis) is reaction between TAG (originated mainly from vegetable oil) and alcohol (usually methanol or ethanol). Methanol has a lower price and favorable physical and chemical properties (non-polar short chain alcohol) and the process of alcoholysis takes place under the mild conditions, with the easier phase separation. Therefore, use of methanol for TAG methanolysis dominate in the process of biodiesel synthesis.

The overall methanolysis reaction occurs as a sequence of three reversible consecutive reactions (Freedman et al., 1986; Noureddini and Zhu, 1997). In the first one, a TAG reacts with methanol to form diacylglycerol and FAME. In the second consecutive reaction a monoacylglycerol and FAME are produced from diacylglycerol and methanol. And finally, in the third reaction sequence glycerol and FAME are formed. This complex reaction could be represented as only one reaction with following stoichiometric equation:

The transesterification reaction can be classified based on the presence and type of the catalyst into non-catalyzed (reaction under supercritical conditions of methanol) and catalyzed process; the later could be divided into chemically and enzyme catalyzed transesterification. Depending on the solubility of the catalyst in the reaction mixture, transesterification can be homogeneously or heterogeneously catalyzed. The selection of biodiesel production process is highly dependent on the feedstock characteristics, mainly the presence of FFA and fatty acids composition of TAG. Main characteristics of each process of biodiesel synthesis are presented in Table 4, and discussed in the following sections.

$$
\begin{array}{llll}
CH_2\text{–}OOCR_1 & & CH_2\text{–}OH & \\
| & & | & \\
CH\text{–}OOCR_2 \;+\; 3\,CH_3OH & \rightleftharpoons & CH\text{–}OH \;+\; 3\,CH_3OOCR_X & (x = 1,2,3) \\
| & & | & \\
CH_2\text{–}OOCR_3 & & CH_2\text{–}OH & \\
\text{TAG} \quad\quad \text{Methanol} & & \text{Glycerol} \quad\quad \text{FAME} &
\end{array}
$$

Table 4. Comparison between different methods for biodiesel production by transesterification

	Homogeneously catalyzed		Supercritical method	Enzyme catalyzed	Heterogeneously catalyzed
	Base	Acid			
Reaction conditions	60–70°C 0.1 MPa	55–80°C 0.1 MPa	239–385°C > 8.09 MPa	20–50°C 0.1 MPa	60–252°C 0.1-24 MPa
Reaction time	1–4 h	1–8h	120–240 s	>10h	0.1–3h
FFA in feedstock	Saponified products	Methyl esters	Methyl esters	Methyl esters	Depend on catalyst, usually not sensitive
Water in feedstock	Interfere with reaction	Interfere with reaction	No influence	Depend on lipase type	Positive up to certain amount
Recovery of glycerol	Difficult	Difficult	Easy	Easy	Easy
Purification of FAME	Repeated washing	Repeated washing	None	None	Easy
Price of catalyst	Low	Low	No catalyst	High	Potentially low

Homogeneously Catalyzed Transesterification Processes

Homogeneously base catalyzed transesterification is the most studied and most commonly used commercial biodiesel production process. This is because the process is realized under the mild reaction conditions (usually close to the boiling temperature of alcohol at ambient pressure), with a lower amount of alcohol (alcohol to oil molar ratio of 6:1 is usually used) and for a short reaction time. A high conversion of TAG (97% or more) could be achieved using most common base catalysts (NaOH, KOH, carbonates and alcoxides: natrium or potassium methoxide, ethoxide or propoxide) (Freedman et al., 1984; Fukuda et al., 2001; Ma and Hanna, 1999; Srivastava and Prasad, 2000; Vicente et al., 2004; Zhang et al., 2003). The highest yield could be achieved using methoxide, because the losses of catalyst as consequence of TAG saponification are negligible since their structures lack the OH group necessary for saponification, but they are expensive and very hygroscopic, which makes them difficult to work with (Vicente et al., 2004). Commercially, the most commonly used are alkaline metal hydroxides (KOH and NaOH) which are much cheaper, but since they are less active than alkoxides, they can give the same high conversions of vegetable oils simply by increasing the catalyst concentration to 1 or 2% (Helwani et al., 2009). The optimal amount of the base catalyst is about 1% (based on oil weight), although slightly lower or higher catalyst concentrations have also been reported.

Beside a lot, above mentioned, advantages, the use of homogeneous catalysts in transesterification process has many drawbacks. Base-catalyzed alcoholysis is very sensitive to the presence of impurities in the oil, primarily to the presence of FFA and water, which is its main disadvantage. The presence of water affects the process of saponification indirectly, via the hydrolysis of TAG, while FFA react directly with alkaline catalyst to form soaps. Both reactions are undesirable side-reactions, that reduce efficiency of the catalyst by its partially consuming and thus lowering its amount in the reaction mixture, leading to decreased biodiesel yield and complicated separation and purification steps. The best yield could be achieved with completely anhydrous reactants and with FFA content lower than 0.5%, although the process still could be catalyzed by alkaline catalyst if FFA content is higher (up to 5%), but catalyst must be added in excess to overcome the saponification problem (Demirbaş et al., 2009a; Ma and Hanna, 1999).

Homogeneous catalyst cannot be reused, and cannot be used in continuous biodiesel production, which is additional disadvantage of homogeneously catalyzed process. Another problem related to the use of a base catalyst is complete removal of these catalysts, which is technically difficult and brings extra cost to the final product (Helwani et al., 2009). In order to meet the specified biodiesel quality, the process involves a number of washing and purification steps. Water washing has been the most frequently used process (Skala et al., 2004), while new purification technologies, beside wet washing (with water or organic solvent) include dry washing (by adsorption or ion-exchange process), membrane separation and ionic liquid treatment (Stojković et al., 2014). Purification processes produce a large amount of wastewater, which is environmentally unfavorable and requires appropriate treatment therefore increasing the overall process cost.

For these reasons, it can be concluded that homogeneously catalyzed alcoholysis show good solution for biodiesel synthesis when vegetable oil with high quality, i.e., refined, edible oil is used. Even then, the purification steps increase process cost, and negatively affect the environment. However, this process is not suitable to produce esters from unrefined oils. Non-edible and waste vegetable oils, generally, contain significant amounts of FFA, as well as water, which limit the use of high effective base catalysts. To overcome the drawbacks of conventional, homogeneously base-catalyzed process alternative methods have been developed. Table 5 summarizes the catalyst type and the optimal reaction conditions for some homogeneously catalyzed (base and acid) alcoholysis of different vegetable oils. Almost in all reported studies where the FFA content in oil was low, the ester yield higher than 90% was achieved, independently of the type of used catalyst and vegetable oil. However, when oil with higher content of FFA was used in transesterification, FAME yield was not satisfactory (Berchmans and Hirata, 2008; Kansedo et al., 2010; Saravanan et al., 2010).

Feedstock	FFA, %	Catalyst	Reaction conditions				Yield, %	Reference
			T, °C	Methanol:oil molar ratio	Wt % of catalyst	Time, h		
Edible oils Soybean oil	0.02	NaOH	20–60	6:1	0.5	1	96	Freedman et al. (1984)
Soybean oil	0.04	NaOH	30–70	6:1	0.2	1.5	63–95	Noureddini and Zhu (1997)
Sunflower oil	0.02	KOH NaOH CH_3ONa	65	6:1	1	0.5 0.75 0.75	~100	Vicente et al. (2004)
Sunflower oil Soybean oil	–	CH_3ONa	60	6:1	0.5	1	98	Freedman et al. (1984)
Rapeseed oil	–	KOH	65	6:1	1	2	96	Rashid et al. (2008)
Canola oil	<0.25	NaOH	40–45	6:1	1	1	93.5	Leung i Guo (2006)
Palm oil	–	KOH	65	6:1	1	2	>96	Darnoko i Cheryan (2000)
Crude palm oil	6.1	NaOH	65	0.28 w/w	1	2	85	Berchmans and Hirata (2008)
Cottonseed oil	–	CH_3ONa	60	6:1	0.5	1	93	Freedman et al. (1984)
Peanut oil	–	CH_3ONa	60	6:1	0.5	1	96	Freedman et al. (1984)
Coconut oil	1.2	NaOH	65	0.28 w/w	1	2	87	Berchmans and Hirata (2008)
Soybean	0.5	H_2SO_4	65	6:1, 20:1, 30:1	1	69	96–98	Freedman et al. (1984)
Non-edible oils *Jatropha curcas*	3.1	NaOH	60	5.6:1	1	1	98	Chitra et al. (2005)
Jatropha curcas	15.0	NaOH	65	0.7 w/w	3.3	2	55	Berchmans and Hirata (2008)
Jatropha curcas	5.53	KOH	50	6:1	1	2	97	Berchmans et al. (2004)
Jatropha curcas	–	CH_3ONa	45	9:1	0.8	0.5	96.3	Tapanes et al. (2008)
Jatropha curcas, two step	21.5 1.0	H_2SO_4 NaOH	65 50	3:7 3:7	1 1	3 3	21.2 90.1	Jain and Sharma (2010)

Table 5. (Continued)

Feedstock	FFA, %	Catalyst	Reaction conditions				Yield, %	Reference
			T, °C	Methanol:oil molar ratio	Wt % of catalyst	Time, h		
Jatropha curcas, two step	15.0 1.0	H_2SO_4 NaOH	50 65	0.6 w/w 0.24 w/w	1 1.4	1 2	90	Berchmans and Hirata (2008)
Karanja (*Pongamia pinnata*)	0.3	KOH	65	6:1	1	2	98	Meher et al. (2006)
Karanja (*Pongamia pinnata*)	–	KOH	60	10:1	1	1.5	92	Karmee and Chadha (2005)
Karanja (*Pongamia pinnata*), two step	9.9 1.3	H_2SO_4 NaOH	55 55	6:1 8:1	1.5 0.8	1 1	95–97	Sharma and Singh (2010)
Mahua (*Madhuca indica*)	14.0	H_2SO_4	65–70	7.5:1	6	5	75	Saravanan et al. (2010)
Mahua (*Madhuca indica*), two step	11.5 1.0	H_2SO_4 NaOH	55 55	6:1 8:1	1.5 0.8	1 1	95–97	Sharma and Singh (2010)
Tobacco seed oil	–	NaOH	55	6:1	1.5	1.5	86	Usta (2005)
Tobacco seed oil, two step	35.0 2.0	H_2SO_4 KOH	60 60	13:1 6:1	2 1	0.8 0.5	24.3 91	Veljković et al. (2006)
Sea mango (*Cerbera odollam*) oil	6.4	NaOH	64.7	6:1	1	1	8.3	Kansedo et al. (2009)
Waste vegetable oils	1.1	NaOH	60	7:1	1.1	0.25	88.8	Leung i Guo (2006)
	0.21	NaOH	65	4.8:1	0.6	1	93	Felizardo et al. (2005)
	–	KOH	65	6:1	1	1	95–96	Refaat et al. (2008)
	37.9	H_2SO_4	95	20:1	4	10	≥90%	Wang et al. (2006)

Homogeneously acid catalyzed transesterification has many drawbacks, such as slower reaction rate, typically 4000 times compared to base catalyzed process, lower catalyst activity, a need for higher reaction temperature and higher alcohol to oil molar ratio. The possibility of simultaneous esterification of FFA and transesterification of TAG when acid is used as a catalyst in the transesterification reaction is important advantage of this process. The tolerance and less sensitivity towards the high FFA presence make acid catalyst suitable for biodiesel production from low-quality feedstock such as non-edible and waste vegetable oils. The most often used acid catalysts are sulfuric acid, phosphoric acid, hydrochloric acid, etc. High biodiesel yield can be achieved in acid-catalyzed transesterification, but a much longer time is needed, even 69 h (Freedman et al., 1984), compared to the base catalyzed process (usually not longer then 1 to 2 h), except at high temperatures (Wang et al., 2006).

Two-Step Process

Due to the disadvantages of both base and acid catalysts, two-step processes for biodiesel production from oils with a high FFA content have been developed. This process consists of acid catalyzed FFA esterification in the first step for reducing the FFA, followed by base catalyzed transesterification of TAG. Pilot plant was developed where in the first step sulfuric acid was used to reduce the FFA content below 1%, after which transesterification using KOH was performed, giving the high yield of FAME (Çanakçi and Van Cerpen, 2003). The only disadvantage of the two-step transesterification process is removal of catalyst in both stages, which causes the higher production cost as compared to conventional, one-step process. In the first step a solid catalyst for esterification of FFA could be used, such as $Fe_2(SO_4)_3$ (Patil et al., 2009; Wang et al., 2007) or HF/SiO_2 (Corro et al., 2011), followed by base catalyzed transesterification of TAG in the second step.

Heterogeneously Catalyzed Transesterification

The heterogeneous catalytic process for biodiesel synthesis is environmentally friendly, since it offers simplified production and product separation and purification processes, with reduced waste water amount. The major advantage of using a solid catalyst is the possibility of their easy regeneration and reuse, which make the biodiesel synthesis process cost-effective and enables the continuous process development (López Granados et al., 2009a; Ma and Hanna, 1999). The synthesis of FAME based on heterogeneous catalytic process could substantially reduce energy consumption (for almost 50%) and thus significantly decrease the total biodiesel production cost (Glišić et al., 2009).

On the other hand, heterogeneously catalyzed process has some disadvantages, first of all the existence of a complex, three-phase system, consisting of a solid (catalyst) and two immiscible liquid phases (oil and methanol), where mass-transfer limitation at the beginning of biodiesel synthesis influences the overall process rate of TAG transesterification. Also, catalyst synthesis procedures are sometimes complicated which lead to higher production cost of biodiesel. Moreover, one of the main problems with solid catalysts is their deactivation with time owing to many possible phenomena, such as poisoning, coking, sintering and leaching (Refaat, 2011). The problem of poisoning is particularly evident when the process involves low-quality oils. More general and dramatic problem is catalyst leaching, which not only can increase the operational cost as a result of replacing the catalyst, but also leads to product contamination, that requires further purifying of biodiesel thereby losing the advantages of heterogeneously catalyzed transesterification processes.

Various compounds were investigated as solid catalysts for biodiesel synthesis, such as alkaline earth metals and other single metals, mixed metal oxides, hydroxides, methoxides, salts of different metals, ion–exchange resins, zeolites, Al–Mg hydrotalcites, etc. Table 6 summarizes the catalyst type and the reaction conditions applied in heterogeneously catalyzed transesterification of different feedstocks. The catalytic activity of a heterogeneous catalyst depends on its nature, specific surface area, pore size and volume, acidity or basicity, catalyst active site concentration and applied reaction conditions (Miladinović et al., 2010; Refaat, 2011). Wide range of temperatures, molar ratio of methanol to oil and catalyst concentration was used in the research of heterogeneously catalyzed transesterification. Temperature ranged from room temperature (Reddy et al., 2006) to as high as 260°C, temperature at which methanol is in the supercritical condition (Wang and Yang, 2007), although the most common applied condition was temperature close to boiling temperature of methanol. Molar ratio of methanol to oil ranged from 4:1 to over 100:1, and catalyst concentration goes to even 40% based on oil, while the time of process duration depends on other conditions, primarily temperature, molar ratio and the type of catalyst.

Regarding the feedstock, base solid catalysts are, as expected, sensitive to the presence of FFA and thus preferable in the transesterification of oil with the lower FFA content, while acid solid catalysts are often used for FAME synthesis from non-edible and waste vegetable oils with the high FFA content. In recent years, due to the fact that low-quality oils become the most important potential feedstock for biodiesel production, research in the field of heterogeneously catalyzed methanolysis is aimed at developing of bifunctional catalyst i.e., catalytic systems that exhibit dual basic and acidic sites which could simultaneously catalyze the transesterification reaction of TAG and FFA esterification in a one-step process.

Two Step Process

Only a few reports are found on the completely heterogeneous two-step process, where in both steps heterogeneous catalyst was used (Srilatha et al., 2012; Zhang et al., 2010). To obtain biodiesel from used cooking oil with 12% of FFA, the free fatty acids were first esterified with methanol using a 25% TPA/Nb_2O_5 catalyst based on oil, followed by transesterification of the oil with methanol over ZnO/Na-Y zeolite catalyst (Srilatha et al., 2012). Beside high yield, both catalysts could be reused several times. The FFA content from *Zanthoxylum bungeanum* seed oil was reduced from 20% to 1% using ferric sulfate, and then transesterification with CaO was performed (Zhang et al., 2010).

Feedstock	FFA, %	Catalyst	Reaction conditions				Yield, %	Reference
			T, °C	Methanol: oil molar ratio	Wt % of catalyst	Time, h		
Edible oils Soybean oil	0.05	CaO	65	10:1	1	2	>99	Kouzu i sar. (2008a)
Soybean oil	–	nano CaO	23-25	27:1	1	24	99	Reddy et al. (2006)
Soybean oil	–	ZnO	130	55:1	5	7	30	Antunes et al. (2008)
Soybean oil	a) – b) 10	Zn glycerolate	140	30:1	3	6	a) 76.6 b) 93.8	Reinoso et al. (2014)
Soybean oil	1.3	ETS-10 zeolite, K-ETS10, Cs-ETS10	150	6:1	10	24	88.5–95.8	Suppes et al. (2004)
Soybean oil	<0.05	Mg–Al hydrotalcite	65	15:1	7.5	9	67	Xie et al. (2006)
Sunflower oil	–	CaO	60 (N2)	13:1	1	1.5	> 90	López Granados et al. (2007)
Sunflower oil	0.14	CaO	60	6:1	1	2	98	Veljković et al. (2009)
Sunflower oil	0.1	$CaO \cdot ZnO$	60	10:1	2	4	97.5	Kesić et al. (2012); Lukić et al. (2014)
Sunflower oil	0.14	$CaTiO_3$, $CaMnO_3$, $Ca_2Fe_2O_5$, $CaZrO_3$	a) 60 b) 165	10:1	2	3	a) <0.5 b) >90	Kesić et al. (2016)
Sunflower oil	0.14	Ca diglyceroxide	60	10:1	0.5	2	97.9	Lukić et al. (2016)
Sunflower oil	–	K_2CO_3/Al-O-Si	120	15:1	2	1	98	Lukić et al. (2009; 2010)
Sunflower oil	0.1	$CaO \cdot ZnO \cdot K_2CO_3$	70	10:1	2	1	>95	Kesić et al. (2015)
Sunflower oil	–	ZnO/Al-O-Si	200	30:1	2	4	92	Lukić et al. (2014)
Rapeseed oil	0.15	Ca/Zr mixed oxide	120	72:1	8	6	92.6	Liu et al. (2015)
Rapeseed oil	<0.5	CaO/MgO	64.5	18:1	2	6	92	Yan et al. (2008)
Canola oil	–	Nano CaO	50–65	9:1	3	2	100	Zhao et al. (2013)
Canola oil	1.04	K_2CO_3/nano CaO	65	9:1	3	8	97.7	Degirmenbasi et al. (2015)
Canola oil	<1	KOH/MgO	65	6:1	3	9	95.1	Ilgen and Akin (2009)
Canola oil		dolomite	67.5	6:1	1.5	3	91.8	Ilgen (2012)
Palm oil	–	CaO/Al_2O_3	65	12:1	3.5	5	95	Zabeti et al. (2009)
Palm oil	–	KOH/Al_2O_3	60	15:1	3	2	91.1	Noiroj et al. (2009)
Palm oil	0.2	SO_4^{2-}/ZrO_2	180	8:1	0.06	2	82.8	Kansedo and Lee (2012)

Table 6. (Continued)

Feedstock	FFA, %	Catalyst	Reaction conditions				Yield, %	Reference
			T, °C	Methanol: oil molar ratio	Wt % of catalyst	Time, h		
Palm kernel oil	–	mixed oxide CaMgZn	60	20:1	6	3	97.5	Limmanee et al. (2008)
Cottonseed oil	–	KF/Mg-La	65	12:1	5	1	~100	Song et al. (2011)
Peanut oil	0.22	dolomite	64	12:1	4	2	96.2 (glycerol yield)	Niu et al. (2014)
Coconut oil	5.8	$Ca(NO_3)_2/Al_2O_3$	60	65:1	15-20	3	94	Benjapornkulaphong et al. (2009)
Coconut oil	0.61	MgZnAlO	182	16:1	3.32	6	98	Olutoye and Hameed (2013)
Corn oil	–	CaO/SiO_2	65	16:1	6	8	90.6	Moradi et al. (2014)
Non-edible oils Jatropha curcas	–	CaO	70	9:1	1.5	2.5	93	Zhu et al. (2006)
Jatropha curcas	7.5	Zr/CaO	35–65	15:1	5	2	>99	Kaur and Ali (2014)
Jatropha curcas	5.53	KNO_3/Al_2O_3	70	12:1	6	6	84	Vyas et al. (2009)
Jatropha curcas	–	CaMgO	65	15:1	4	5	83	Taufiq-Yap et al. (2011)
Jatropha curcas	–	Mg–Zr mixed oxid	65	55:1	10	0.75	~100	Sree et al. (2009)
Jatropha curcas	7.23	MgZnAlO	182	11:1	3.32	6	92	Olutoye and Hameed (2011)
Karanja (Pongamia pinnata)	3.4	Li/CaO	65	12:1	5	2	99	Kaur and Ali (2011)
Karanja (Pongamia pinnata)	–	a) ZnO b) Montmorillonite K-10 c) Hβ-Zeolite	120	10:1	24	0.1	a) 83 b) 47 c) 59	Karmee and Chadha (2005)
Castor (Ricinus communis) oil	–	hydrated lime	a) 25 b) 65	0.42:1 vol%	1.6	a) 14 b) 2	a) 98 b) 100	Sánchez-Cantú et al. (2013)
Castor (Ricinus communis) oil	–	TiO_2/SO_4	120	6:1	5 mol%	1	25	Almeida et al. (2008)
Castor (Ricinus communis) oil	–	$Zn_5(OH)_8(NO_3)_2{\cdot}2H_2O$	60	29:1	5	3	20	Zięba et al. (2010)
Sea mango (Cerbera odollam) oil	6.4	SO_4^{2-}/ZrO_2	150	12:1	0.08	3	94.1	Kansedo and Lee (2012)

Feedstock	FFA, %	Catalyst	Reaction conditions				Yield, %	Reference
			T, °C	Methanol: oil molar ratio	Wt % of catalyst	Time, h		
Waste vegetable oil	3.7	CaO	65	12:1	5	4	86–87	Puna et al. (2013)
	3.66	CaO	65	12:1	5	4	87	Soares Dias i sar. (2013)
	2.6	CaO	65	10:1	1	1	>99	Kouzu i sar. (2008a)
	0.87	$CaO–ZrO_2$	65	30:1	10	2	92.1	Dehkordi and Ghasemi (2012)
	3.78	$CaO–La_2O_3$	58	20:1	5	3	94.3	Yan et al. (2009)
	0.45	$Ca_{0.9}Mn_{0.1}O$	50	18:1	3	4	94.9	Dias et al. (2013)
	1.27	CaO·ZnO	60–96	10:1	2	0.25–1	98	Lukić et al. (2013)
	1.35	MgZnAlO	182	16:1	9	6	85	Olutoye i Hameed (2013)
	2.54	SO_4^{2-}/SnO_2	150	15:1	3	3	92.3	Lam and Lee (2010)
	1.47	Ti–SBA–15	200	108:1	15	3	97.6	Chen et al. (2014)
	0.52	SrO	65	6:1	5	3	86	Viola et al. (2012)
	15	Zn stearat/ SiO_2	200	18:1	3	10	98	Jacobson et al. (2008)
	15	Quntinite-3T	75	15:1	10	6	97.7	Kondamudi et al. (2011)

Enzyme-Catalyzed Transesterification Processes

Synthesis of biodiesel using enzyme catalyst is a green approach of producing renewable fuel by environmental benign biocatalyst. Enzyme-catalyzed transesterification process has certain advantages compared to the chemically catalyzed transesterification process s such as mild reaction conditions, simultaneously catalysis of FFA esterification and TG transesterification which make this process suitable for feedstock with high FFA content, no difficulty in the enzyme removal from reaction mixture and product separation and purification. However, the major issue for the industrialization of the enzyme-catalyzed biodiesel production is the high price of the enzyme. One way to overcome this drawback is to immobilize the enzyme and obtain a reusable and more stable catalyst for the continuous processes. Lipases from different origins, such as *Pseudomonas fluorescens*, *Pseudomonas cepacia*, *Rhizomucor miehei*, *Rhizopus oryzae*, *Candida rugosa*, *Chromobacterium viscosum*, *Thermomyces lanuginosus*, *Candida antarctica* etc. are most commonly used as catalyst for biodiesel production (Bajaj et al., 2010; Banković-Ilić et al., 2012; Guldhe et al., 2015). Celite, macroporous acrylic resin, macroporous anion exchange resin, silica and reticulated polyurethane foam are some of the carriers used for enzymes immobilization (Banković-Ilić et al., 2012). Recently, novel immobilization techniques are being developed to enhance the performance of immobilized lipase, solvent tolerance, reusability, stability, and to make the separation process easier (Guldhe et al., 2015). These techniques include the new types of carriers such as protein-coated microcrystals, cross-linked protein coated microcrystals, magnetic particle carriers and different nanofibrous membranes such as hydrophobic polymer and polyvinylidene fluoride membranes (Guldhe et al., 2015). Examples of enzyme-catalyzed transesterification of different feedstocks, with enzyme type, carrier, solvent and the reaction conditions applied are given in Table 7.

Beside the high cost of enzymes, the problems related to the use of enzymes as catalysts in transesterification process are much longer reaction time and inhibition of lipase activity by short chain alcohol. Methanol can deactivate immobilized lipases, thus the use of ethanol, propan-2-ol, ethyl acetate or diethyl carbonate as acyl acceptors has been suggested to overcome lipase inactivation (Banković-Ilić et al., 2012). The other solution is the stepwise addition of methanol. The addition of organic solvents such as *n*-hexane, *t*-butanol, *n*-heptan, 1,4-dioxane, petroleum ether and iso-octane or the enzyme-catalyzed transesterification under supercritical CO_2 improve the mass transfer and reaction rate (Banković-Ilić et al., 2012; Guldhe et al., 2015).

The biodiesel yield can be enhanced and the time of reaction reduced by using two immobilized lipases from different sources which have different substrate specificity and catalytic activity (Bajaj et al., 2010; Guldhe et al., 2015). The combination of two immobilized enzymes, of which one is active in the catalysis of esterification of FFA and the other in the transesterification of TAG, enables biodiesel synthesis from low-cost feedstock.

The enzyme-catalyzed process is generally performed at relatively low temperatures, between 20 and 50°C, because at higher temperatures denaturation of enzyme occurs, thereby decreasing its activity (Guldhe et al., 2015). Water is a very important factor, because it influences the lipase catalytic activity and stability. Excess water leads to ester hydrolysis, which decreases the esters yield. Lipases from different sources showed different responses towards water content, thus optimum water content in the reaction depends upon the lipase and feedstock used, immobilization technique employed and type of solvent (Guldhe et al., 2015).

Supercritical Transesterification Process

Transesterification can take place without the presence of a catalyst and is carried out with alcohols at temperatures and pressures above their critical values i.e., at supercritical conditions. Under the condition of supercritical state thermo-physical properties such as dielectric constant, viscosity, specific gravity and polarity, drastically change depending on the temperature and pressure. These properties of densed, supercritical alcohol enable better solubility of TAG and alcohol. One of the most important alcohol characteristics, its polarity, would decrease in supercritical conditions, and as a result, non-polar TAG can be well dissolved with supercritical alcohol to form a single phase of oil/alcohol mixture (Kusdiana and Saka, 2001; 2004a). In this way, the initial reaction lag stage caused by the low solubility of the alcohol in the oil phase can be overcome, the reaction rate is higher, the reaction time is shorter, and in several minutes high yield of biodiesel can be achieved.

One of the main advantages of supercritical transesterification is the absence of catalyst, which makes separation and purification of reaction products, since there is no soap formation and no waste generation from catalyst neutralization and purification of biodiesel. Another very important feature of supercritical transesterification is that water and FFA do not have any negative effect on the conversion, therefore this method of biodiesel production can be used for various low-grade biodiesel feedstocks. FFA are easily converted into corresponding esters and thus contribute to higher FAME yields. If water exists in reaction system, hydrolysis of TAG occurs, after which esterification between formed FFA and methanol takes place, thus, the presence of water at a certain amount in transesterification could even enhance the methyl esters formation (Kusdiana and Saka, 2004a).

The main disadvantages of supercritical reaction are extremely high temperature (about 350°C) and pressure (45 MPa) which lead to the high energy consumption and high cost of the apparatus since the expensive materials should be used for making reaction equipment to avoid corrosion and ensure the safety (Lee and Saka, 2010). High temperatures also can cause the easy degradation of produced esters. Also, the large amount of alcohol is used in supercritical transesterification, usually molar ratio of methanol to oil is about 40:1. The review of the operating conditions applied in supercritical processes is given in Table 8.

To overcome these disadvantages, Kusdiana and Saka (2004b) developed a two-step biodiesel production method (Saka–Dadan process) in which hydrolysis of TAG is carried out at a subcritical state of water to obtain fatty acids, while the methyl esterification of the hydrolyzed products is treated near the supercritical methanol condition to obtain FAME. The two-step method use milder reaction condition and shorter reaction time, as an advantage compared to one step process.

Saka and Isayama (2009) proposed a non-catalytic biodiesel production process by using supercritical methyl acetate instead of alcohol, thus producing the FAME and triacetin instead of glycerol. Triacetin formed as side product can also be used in the cosmetic, food or petrochemical industries. It was found that the reactivity of methyl acetate is slightly lower than methanol in supercritical state, and transesterification can only proceed at a temperature higher than 320°C.

Table 7. Examples of enzyme-catalyzed transesterification of different feedstocks

Feedstock	FFA, %	Enzyme	Carrier	Solvent/ water	Reaction conditions				Yield, %	Reference
					T, °C	Alcohol:oil molar ratio	Wt % of catalyst	Time, h		
Edible oils Soybean oil	–	*Candida rugosa*	Magnetic chitosan microsphere	*n*–hexane	35	4:1 (methanol stepwise addition)	10	30	87	Xie and Wang (2012)
Soybean oil	–	*Thermomyces lanuginosa*	Silica gel	–	40	1:1 (methanol stepwise addition)	10	50	>90	Du et al. (2005)
Sunflower oil	–	a) R*hizomucor miehei* b) *Thermomyces lanuginosa* c) *Pseudomonas fluorescens*	a)Ion exchange resin b) Silica gel c) Polypropylene	*n*–hexane	40	Methanol 3:1	10	48	a) 80 b) 85 c) 91	Soumanou and Bornscheuer (2003)
Sunflower oil	–	*Candida antarctica* (Novozyme 435)	Acrylic resin	–	45	Methanol 3:1	3	1	>99	Ognjanović et al. (2009)
Sunflower oil	–	*Candida antarctica* (Novozyme 435)	Macroporous acrylic resin	–	50	Propan-2-ol 4:1	10	8	93.4	Modi et al. (2006)
Rapeseed oil	1.8	3% Lipozyme TL IM (immobilized *Thermomyces lanuginosa*) + 1% Novozym 435 (immobilized *Candida antarctica*)	–	*t*-butanol	35	Methanol 4:1	4	12	95	Li et al. (2006)
Canola oil	–	a) *Candida antarctica* b) *Thermomyces lanuginosus* c) *Rhizomucor miehei*	Epoxy functionalized SBA-15	1) – 2) *t*-butanol	50	Methanol 3:1	3.8	72	1) a) 24 b) 33 c) 34 2) a) 58.8 b) 94.2 c) 95.4	Babaki et al. (2016)
Palm oil	–	*Thermomyces lanuginosus*	Modified Fe_3O_4	*t*-butanol	50	Methanol to FFA 4.7:1	23.2	24	97.2	Raita et al. (2015a)
Crude palm oil	4	*Thermomyces lanuginosus*	cross-linked protein coated microcrystalline+ zwitterionic glycine	*t*-butanol	50	Methanol to FFA 4:1	20	60	95.5	Raita et al. (2015b)
Cottonseed oil	–	*Candida antarctica*	Macroporous resin	*t*-butanol	50	Methanol 6:1	1.7	24	97	Royon et al. (2007)
Coconut oil	1.2	Novozym 435	–	–	60	Ethanol 10:1	7	24	80.5	Berchmans and Hirata (2008)

Feedstock	FFA, %	Enzyme	Carrier	Solvent/ water	Reaction conditions				Yield, %	Reference
					T, °C	Alcohol:oil molar ratio	Wt % of catalyst	Time, h		
Non-edible oils *Jatropha curcas*	–	*Rhizopus oryzae*	Reticulated polyurethane foam particles	–	30	Methanol 1:1	4	60	80	Tamalampudi et al. (2008)
Jatropha curcas	2.71	*Pseudomonas cepacia*	Celite	Water 4-5%	50	Ethanol 1:4	10	8	98	Shah and Gupta (2007)
Jatropha curcas	4.65	*Burkholderia cepacia*	Modified attapulgite	Water 7%	35	Methanol 6.6:1	10	24	94	You et al. (2013)
Jatropha curcas	–	*Candida antarctica (Novozyme 435)*	Macroporous acrylic resin	–	50	Propan-2-ol 4:1	10	8	92.8	Modi et al. (2006)
Jatropha curcas	–	*Candida antarctica (Novozyme 435)*	Macroporous acrylic resin	–	50	Ethyl acetate 11:1	10	11	91.3	Modi et al. (2007)
Karanja (*Pongamia pinnata*)	–	*Candida antarctica (Novozyme 435)*	Macroporous acrylic resin	–	50	Propan-2-ol 4:1	10	8	91.7	Modi et al. (2006)
Karanja (*Pongamia pinnata*)	–	*Candida antarctica (Novozyme 435)*	Macroporous acrylic resin	–	50	Ethyl acetate 11:1	10	11	90.0	Modi et al. (2007)
Mahua (*Madhuca indica*)	–	*Pseudomonas cepacia*	a) K_2SO_4 microcrystals b) Glutaraldehyde c) Accurel	–	40	Ethanol 4:1	10	a) 2.5 b) 2.5 c) 6	a) 99 b) 92 c) 96	Kumari et al. (2007)
Waste vegetable oils	18.3	*Geotrichum* sp.	a) K_2SO_4 microcrystals b) Glutaraldehyde	*t*-butanol	40	Methanol 4:1	0.5	a) 9 b) 12	a) 72 b) 69	Yan et al. (2011)
	70	3% Lipozyme TL IM (immobilized *Thermomyces lanuginosa*) + 1% Novozym 435 (immobilized *Candida antarctica*)	–	*t*-butanol	35	Methanol 4:1	4	5	>75	Li et al. (2006)
	27.2	*Penicillium expansum*	Resin D4020	*t*-amyl alcohol + blue silica gel as adsorbent	35	Methanol 3:1	168 U	7	92.8	Li et al. (2009)

Table 8. Examples of supercritical transesterification of different feedstocks

Feedstock	FFA, %	Reaction conditions					Yield, %	Reference
		T, °C	Pressure, MPa	Methanol:oil molar ratio	Catalyst/ wt %	Time, min		
Edible oils Soybean oil	–	310	32	40:1	–	residence time: 25	96	He et al. (2007)
Soybean oil	–	260	30	36:1	nano MgO/3	10	~100	Wang and Yang (2007)
Sunflower oil	–	400	20	50:1	–	30	96	Madras et al. (2004)
Sunflower oil	–	252	24	41:1	CaO/3	6	~100	Demirbaş (2007b)
Rapeseed oil	–	350	45	42:1	–	4	95	Saka and Kusdiana (2001)
Rapeseed oil	1.5	300	20	42:1	–	15	~100	Warabi et al. (2004)
Rapeseed oil	–	350	20	42:1 methyl acetate	–	45	97	Saka and Isayama (2009)
Rapeseed oil	–	350	20	42:1 dimethyl carbonat	–	12	94	Ilham and Saka (2009)
Palm oil	–	400	20	50:1	–	30	96	Rathore and Madras (2007)
Palm kernel oil	15.5	350	19	42:1	–	space time: ~7	96	Bunyakiat et al. (2006)
Cottonseed oil	–	250	N/A	41:1	–	8	98	Demirbaş (2002)
Hazelnut kernel oil	–	250	N/A	41:1	–	5	95	Demirbaş (2002)
Coconut oil	8	350	19	42:1	–	space time: ~7	95	Bunyakiat et al. (2006)
Linseed oil	–	350	20	40:1	–	30	~100	Varma and Madras (2007)
Non-edible oils *Jatropha curcas*	–	400	20	50:1	–	30	96	Rathore and Madras (2007)
Jatropha curcas	2	320	8.4	43:1	–	4	100	Hawash et al. (2009)
Jatropha curcas	3.5	320	15	40:1	–	5	84.6	Samniang et al. (2014)
Jatropha curcas	0.46	250	7	24:1	Micro NaOH/0.8	28	90.5	Tang et al. (2007)
Karanja (*Pongamia pinnata*)	–	400	20	50:1	–	20	96	Rathore and Madras (2007)
Karanja (*Pongamia pinnata*)	2.2	300	20	42:1 methyl acetate	–	45	96.6	Goembira and Saka (2015)
Castor (*Ricinus communis*) oil	–	350	20	40:1	–	30	~100	Varma and Madras (2007)
Sea mango (*Cerbera odollam*)	6	380	8–10	45:1	–	40	78	Ang et al. (2015)
Waste vegetable oils	–	360	25	40:1	–	20	80	Tan et al. (2011)
	4	270	10	2:1	–	45	100	Lee et al. (2012)
	1.32	287	N/A	41:1	–	30	99.6	Demirbaş et al. (2009b)

Demirbaş, 2009; Gui et al., 2008; Karmakar et al., 2010; Karmee and Chadha, 2005; Leung et al., 2010; Ramos et al., 2009; Srivastava and Prasad, 2000; Talebian-Kiakalaieh et al., 2013)

Vegetable oil	Kinematic viscosity at 40°C, mm^2/s	Density, kg/m^3	Calorific value, MJ/kg	Cetan number	Flash Point, °C	Pour point, °C	Cloud point, °C	Oxidation stability at 110°C, h
Edible oils Soybean	4.04–4.5	885–914	40–41.3	45–52	178	–7 to –3	–2 to 1	1.3–2.1
Sunflower	4.2–4.9	860–880	41.3–45.3	49–50	166–183	–	1–3.4	0.8–0.9
Rapeseed	4.3–5.8	882	37–45	54.4–55	170	–12	–4 to –3.3	2.0–7.6
Palm	4.4–5.7	860–900	34–41.2	61–62	164–182	15	13–15	4.0
Canola	3.53	880–900	–	56	–	–	–	–
Linseed	3.75	892.5	37	53	172	–	–3.8	0.2
Coconut	2.73	807	–	–	110	–	0.0	35.5
Peanut	4.42–4.9	883	39.7–40.1	54	176–193		5	–
Cottonseed	3.6–9.6	850–880	37.5–41.1	45.5–54	162–167	–4	–2	–
Non-edible oils *Jatropha curcas*	4.23–4.8	863–880	39.2–42.7	52.3–63	135–148	4.2	2.7–10.2	2.3
Karanja (*Pongamia pinnata*)	4.85	876–890	35.6–42	52–61	150–180	–6	–1	2.54
Castor (*Ricinus communis*)	15.25	899	–	–	–	–	–13.4	1.1
Mahua (*Madhuca indica*)	3.98–5.72	880–916	37–39.4	56.6	129–208	6	5	–
Neem (*Azadirachta indica*)	5.2–8.8	820–884.5	40.1	51–57.8	–	–	14.4	7.1
Rubber seed (*Hevea brasiliensis*)	5.8	860	36.5	37–49	130	–8	4	–
Tobacco seed (*Nicotiana tabacum*)	4.23–5.2	860–888	44.6	51.6	165.4	–	–	0.8
Waste vegetable oil	4–5.3	878–897	39.2	48–54	148–196	9–11	13	–
Standard EN14214	3.5–5.0	860–900	35	51	>101	–	–	>8
Petroleum diesel ASTM D975	2.0–4.5	820–860	42–46	46	60–80	–35 to –15	–15 to 5	–

Dimethyl carbonate was also proposed as a potential reactant for biodiesel synthesis under supercritical conditions (Ilham and Saka, 2009). Final products of this process are FAME with glycerol carbonate and citramalic acid as byproducts (Ilham and Saka, 2009). It is reported that glycerol carbonate has higher economic value than the abundantly available glycerol because it can be used as the raw materials for paint, dyes and adhesives and as a source of new polymeric materials (Ilham and Saka, 2009).

Properties and Quality of Biodiesel Obtained from Different Oils

Biodiesel that is used in diesel engines in pure form (B100) must be in accordance with certain standards for similar fuels produced from petroleum. The characteristics of biodiesel, which is made in the process of alcoholysis of vegetable oils, are defined by the specification EN 14214 of the European Committee for Standardization, which completely defines the quality of FAME. The standard was adopted in 2003, has been changed over the years with last version from 2012, and is valid for all member states of the European Union. While EN 14214 is more restrictive and applies only to mono-alkyl esters made with methanol, i.e., fatty acid methyl esters (FAME), the US specification, ASTM D6751, defines biodiesel as mono-alkyl esters of long chain fatty acids derived from vegetable oils and animal fats.

The properties of biodiesel can be influenced by various factors including the quality and physicochemical properties of feedstock, primarily its fatty acid composition, as well as the production method (Atabani et al., 2013; Maddikeri et al., 2012). Fatty acid composition of oil plays an important role in some critical parameters of the biodiesel, such as cetane number, oxidation stability and cold flow properties. Table 9 shows some of the important physical and chemical properties of biodiesel produced from different oil sources as compared to diesel fuel and EN 14214 specification.

Viscosity is very important property of biodiesel since it affects the performance of the fuel flow system and the operation of fuel injection equipment, particularly at low temperatures when the increase in viscosity affects the fluidity of the fuel (Demirbaş, 2005; 2009a). If the viscosity of the biodiesel is lower, it is easier to pump and atomize to achieve finer droplets (Demirbaş, 2005; 2009a). The conversion of TAG into FAME through the transesterification process reduces the molecular weight to one third that of the TAG and reduces the viscosity significantly, from 27.2 and 53.6 mm^2/s for different oils, to 3.59 and 4.63 mm^2/s for FAME, but compared to diesel fuel, all of the vegetable oil methyl esters are slightly viscous. Structural features, fatty acid composition such as chain length, degree of unsaturation, double bond orientation, and type of ester head group can influence the kinematic viscosities of biodiesel (Atabani et al., 2013; Demirbaş, 2009a). Viscosity increases with the increase in the chain length and decreases with the increase in the number of double bonds (unsaturation chain level). As can be seen from Table 9, most of the biodiesel obtained from different oils has acceptable viscosity, with exception of coconut biodiesel that has lower value, due to the high amount of short chain fatty acids, and castor with much higher viscosity due to the high content of ricinoleic acid.

Cetane number (CN) is a measure of ignition quality of diesel fuel and it relates to the ignition delay time of a fuel upon injection into the combustion chamber. The higher the CN, the easier the fuel ignites when it is injected into the engine (Demirbaş, 2005). The CN of biodiesel is generally higher than CN of conventional diesel, because of its higher oxygen content (10 to 11% by weight), which may encourage more combustion than hydrocarbon-based diesel fuels in an engine. Biodiesel CN depends on the feedstock used for its production and increases with increasing chain length of fatty acids and increasing saturation (Karmakar et al., 2010). Lower cetane numbers have been associated with more highly unsaturated components (C18:2 and C18:3). From that aspect, the oils such as sunflower, soybean, rubber seed and linseed, with higher levels of unsaturated components, could hardly meet the standard regarding CN (Ramos et al., 2009). In the case of palm biodiesel the CN is very high because of the presence of saturated fatty acids such as palmitic (C16:0) and stearic (C18:0) acid. Rapeseed oil is rich in C18:1, C20:1, C22:1 and biodiesel produced from this oil has a CN close to palm biodiesel (Table 9). For biodiesel from oils which contain a higher amount of unsaturated linoleic acid (C18:2), such as peanut and corn biodiesels, the cetane number is in the medium range (Ramos et al., 2009). Jatropha, mahua, neem and tobacco have CN close to that of diesel fuel, while karanja usually have higher CN than other biodiesels (Table 9).

Oxidation stability is one of the major issues affecting the use of biodiesel and it determines resistance to chemical changes brought about by oxidation reaction (Ramos et al., 2009). Oxidation stability of biodiesel depends greatly on fatty acid compositions and degree of unsaturation. Saturated FAME are more stable than unsaturated, while polyunsaturated FAME are at least twice as reactive to autooxidation than monounsaturated. Therefore, the biodiesel produced from feedstocks with higher concentrations of saturated fatty acids commonly shows better oxidation stability. Contrary, the vegetable oils rich in linoleic and linolenic acids, such as soybean and sunflower tend to give methyl ester fuels with poor oxidation stability (Table 9). Furthermore, the vegetable oil with presence of smaller amount of polyunsaturated fatty acids, such as palm and neem oil, give methyl esters with improved oxidation stability (Ramos et al., 2009). Coconut oil is high in saturated fatty acids and gives biodiesel with very high oxidation stability. According to these results, it might be stated that an inverse relationship between a high oxidative stability and better cold flow properties of the biodiesel fuel exists (Singh et al., 2014).

Important parameters for low-temperature applications of a fuel are cloud point (CP), pour point (PP) and cold filter plugging point (CFPP). The CP is the temperature at which a cloud of wax crystals first becomes visible when the fuel is cooled (Demirbaş, 2009a). The PP is the temperature at which the amount of wax from a solution is sufficient to gel the fuel; thus it is the lowest temperature at which the fuel can flow (Demirbaş, 2009a). The CFPP defines the fuels limit of filterability and refers to the lowest temperature at which biodiesel flows under vacuum condition through a standard wire mesh filter. The value of CFPP is defined in national standards and ranges from +5 to –26 °C, depending on the season and the region in which adopted standard is applied. The low temperature properties of biodiesel are also mainly dependent on its compositions. It is well known that unsaturated FAME crystallizes at lower temperature than saturated, due to its different three-dimensional configuration. A higher content of unsaturated fatty acids in the vegetable oil will result in biodiesel having better cold flow properties. For example, biodiesel from palm oil has poor cold flow properties because it has high content of saturated fatty acids (about 50%) (Gui et al., 2008). Meanwhile biodiesel from rapeseed oil which has high content of unsaturated fatty acids has the best cold flow properties. The biodiesel CP is typically higher than CP of diesel fuel and ranges from –13 to 15°C. The saturated fatty acids produce methyl esters that will start to crystallize at ~0°C, such as for soybean and sunflower oil. The castor methyl ester has the lowest CP of –13.4°C, while the highest CP are noticed for biodiesel produced from palm, neem and waste cooking oils (13–15°C).

CONCLUSION

The need to replace fossil fuels with alternative and renewable energy sources, partly due to their limited resources and partly due to the increased environmental concern, has led to extensive research in the R&D of biodiesel synthesis and production. The wide range of available feedstock for biodiesel production is one of its advantages, and since the cost of feedstock represents over 70% of overall biodiesel production cost, selecting the best feedstock is vital to ensure the low production cost of biodiesel. The quality and physicochemical properties of feedstock strongly influence the quality of biodiesel, as well as the selection of suitable production technology. Currently, biodiesel is mostly produced from edible oils, however from economic and social reasons, the edible oils should be substituted with lower-cost and reliable feedstock, which do not compete with food crops, such as non-edible and waste vegetable oils. Since the amounts of oily crops, both edible and non-edible, animal fats and waste cooking oils are limited, and unlikely to provide worldwide biodiesel production demand, in recent years a high interest has arisen towards the other potential renewable sources, primarily microalgae.

Increasing biodiesel consumption requires optimized production processes allowing high production capacities, simplified operations, high yields and the absence of special chemical requirements and wastes as side products during FAME synthesis. Conventional homogeneous process of FAME synthesis today is mainly applied in industry, however it has some economical and environmental drawbacks, such as removal of the catalysts after reaction and a large amount of wastewater produced during the catalyst neutralization and FAME washing. What is more important, this method is not suitable for the biodiesel production from oils having higher amount of FFA, which both non-edible and waste vegetable oils in most cases are. The utilization of an efficient solid catalyst could solve these problems of a classical, homogeneous method. The advantage of heterogeneous catalyst usage is its fast and easy separation from the reaction mixture without requiring the use of neutralization agent. Furthermore such type of catalyst could be regenerated and reused, and is suitable for use in continuous process, leading to safer, cheaper and more environment-friendly operation. Another promising technology for biodiesel production from low-cost feedstock is non-catalytic process, i.e., process under supercritical conditions of alcohol. This process is conducted without using any catalyst, avoiding the complications of purification in the downstream processing, and is not sensitive to the presence of FFA and water in the oil.

The main obstacle in biodiesel commercialization is still its high cost, thus it needs to be reduced either by the reduction of feedstock cost, or by the application of more efficient technology for biodiesel production, which at the same time has to be environmentally acceptable. By using a low-cost feedstock, cost of biodiesel can be significantly reduced. Technologies based on utilization of heterogeneous catalyst or the non-catalytic supercritical process have shown high potential in overcoming existing limitations in biodiesel synthesis, however further research is needed for their practical applications.

Acknowledgments

Financial support of this work to Serbian Ministry of Education, Science and Technological Development (Project III 45001) is acknowledged.

References

Almeida, R. M. D.; Noda, L. K.; Goncalves, N. S.; Meneghetti, S. M. P.; Meneghetti, M. R. Transesterification reaction of vegetable oils, using superacid sulfated TiO_2-base catalysts. *Appl. Catal., A* 2008, *347*, 100–105.

Ang, G. T.; Ooi, S. N.; Tan, K. T.; Lee, K.T; Mohamed, A.R. Optimization and kinetic studies of sea mango (*Cerbera odollam*) oil for biodiesel production via supercritical reaction. *Energy Convers. Manage.* 2015, *99*, 242–251.

Antunes, W. M.; de Oliveira Veloso, C.; Assumpção Henriques, C. Transesterification of soybean oil with methanol catalyzed by basic solids. *Catal. Today* 2008, *133–135*, 548–554.

Aransiola, E. F.; Ojumu, T. V.; Oyekola, O. O.; Madzimbamuto, T. F.; Ikhu-Omoregbe, D. I. O. A review of current technology for biodiesel production: State of the art. *Biomass Bioenergy* 2014, *61*, 276–297.

Atabani, A. E.; Silitonga, A. S.; Badruddin, I. A.; Mahlia, T. M. I.; Masjuki, H. H.; Mekhilef, S. A comprehensive review on biodiesel as an alternative energy resource and its characteristics. *Renewable Sustainable Energy Rev.* 2012, *16*, 2070–2093.

Atabani, A. E.; Silitonga, A. S.; Ong, H. C.; Mahlia, T. M. I.; Masjuki, H. H.; Badruddin, I. A.; Fayaz, H. Non–edible vegetable oils: A critical evaluation of oil extraction, fatty acid compositions, biodiesel production, characteristics, engine performance and emissions production. *Renewable Sustainable Energy Rev.* 2013, *18*, 211–245.

Babaki, M.; Yousefi, M.; Habibia, Z.; Mohammadi, M.; Yousefi, P.; Mohammadi, J.; Brask, J. Enzymatic production of biodiesel using lipases immobilized on silica nanoparticles as highly reusable biocatalysts: effect of water, t-butanol and blue silica gel contents. *Renewable Energy* 2016, *91*, 196–206.

Balalić, I.; Zorić, M.; Branković, G.; Terzić, S.; Crnobarac, J. Interpretation of hybrid × sowing date interaction for oil content and oil yield in sunflower. *Field Crops Res.* 2012, *137*, 70–77.

Balat, M. Potential alternatives to edible oils for biodiesel production – a review. *Energy Convers. Manage.* 2011, *52*, 1479–1492.

Banković-Ilić, I. B.; Stamenković, O. S.; Veljković, V. B.; Biodiesel production from non-edible plant oils, *Renewable Sustainable Energy Rev.* 2012, *16*, 3621– 3647.

Benjapornkulaphong, S.; Ngamcharussrivichai, C.; Bunyakiat, K., Al_2O_3-supported alkali and alkali earth metal oxides for transesterification of palm kernel oil and coconut oil. *Chem. Eng. J.* 2009, *145*, 468–474.

Berchmans, H. J.; Hirata, S. Biodiesel production from crude *Jatropha curcas L.* seed oil with a high content of free fatty acids. *Bioresour. Technol.* 2008, *99*, 1716–1721.

Berchmans, H. J.; Morishita, K.; Takarada, T. Kinetic study of hydroxide-catalyzed methanolysis of *Jatropha curcas*–waste food oil mixture for biodiesel production. *Fuel* 2010, *104*, 46–52.

Bunyakiat, K.; Makmee, S.; Sawangkeaw, R.; Ngamprasertsith, S. Continuous Production of Biodiesel via Transesterification from Vegetable Oils in Supercritical Methanol. *Energy Fuels* 2006, *20*, 812–817.

Çanakçi, M. The potential of restaurant waste lipids as biodiesel feedstock. *Bioresour. Technol.* 2007, *98*, 183–190.

Çanakçi, M.; Van Gerpen, J. A pilot plant to produce biodiesel from high free fatty acid feedstocks. *Trans. ASAE* 2003, *46*, 945–954.

Chavanne, G. Procedure for the transesterification of vegetable oils for their uses as fuels. Belgian Patent No. 422877, 1937.

Chen, S.; Mochizuki, T.; Abe, Y.; Toba, M.; Yoshimura, Y. Ti-incorporated SBA-15 mesoporous silica as an efficient and robust Lewis solid acid catalyst for the production of high-quality biodiesel fuels. *Appl. Catal., B* 2014, *148–149*, 344–356.

Chitra, P.; Venkatachalam, P.; Sampathrajan, A. Optimisation of experimental conditions for biodiesel production from alkali-catalysed transesterification of Jatropha curcus oil. *Energy Sustainable Dev.* 2005, *9*, 13–8.

Chhetri, A. B.; Watts, K. C.; Islam, M. R. Waste Cooking Oil as an Alternate Feedstock for Biodiesel Production. *Energies* 2008, *1*, 3–18.

Corro, G.; Tellez, N.; Jimenez, T.; Tapia, A.; Banuelos, F.; Vazquez-Cuchillo, O. Biodiesel from waste frying oil. Two step process using acidified SiO_2 for esterification step. *Catal. Today* 2011, *166*, 116–122.

Darnoko, D.; Cheryan, M.; Kinetics of palm oil transesterification in a batch reactor. *J. Am. Oil Chem. Soc.* 2000, *77*, 1263–1267.

Degirmenbasi, N.; Coskun, S.; Boz, N.; Kalyon, D.M. Biodiesel synthesis from canola oil via heterogeneous catalysis using functionalized CaO nanoparticles. *Fuel* 2015, *153*, 620–627.

Dehkordi, A. M.; Ghasemi, M. Transesterification of waste cooking oil to biodiesel using Ca and Zr mixed oxides as heterogeneous base catalysts. *Fuel Process. Technol.* 2012, *97*, 45–51.

Demirbaş, A. Biodiesel from vegetable oils via transesterification in supercritical methanol. *Energy Convers. Manage.* 2002, *43*, 2349–2356.

Demirbaş, A. Biodiesel fuels from vegetable oils via catalytic and non-catalyticsupercritical alcohol transesterifications and other methods: A survey. *Energy Convers. Manage.* 2003, *44*, 2093–2109.

Demirbaş, A. Biodiesel production from vegetable oils via catalytic and non-catalytic supercritical methanol transesterification methods. *Prog. Energy Combust. Sci.* 2005, *31*, 466–487.

Demirbaş, A. Importance of biodiesel as transportation fuel. *Energy Policy* 2007a, *35*, 4661–4670.

Demirbaş, A. Biodiesel from sunflower oil in supercritical methanol with calcium oxide. *Energy Convers. Manage.* 2007b, *48*, 937–941.

Demirbaş, A. Studies on cottonseed oil biodiesel prepared in non-catalytic SCF conditions. *Bioresour. Technol.* 2008, *99*, 1125–1130.

Demirbaş, A. Progress and recent trends in biodiesel fuels. *Energy Convers. Manage.* 2009a, *50*, 14–34.

Demirbaş, A. Biodiesel from waste cooking oil via base-catalytic and supercritical methanol transesterification. *Energy Convers. Manage.* 2009b, *50*, 923–927.

Dias, J. M.; Alvim-Ferraz, M. C. M.; Almeida, M. F.; Díaz, J. D. M.; Polo, M. S.; Utrilla, J. R. Biodiesel production using calcium manganese oxide as catalystand different raw materials. *Energy Convers. Manage.* 2013, *65*, 647–653.

Du, W.; Xu, Y.Y.; Liu, D. H.; Li, Z. B. Study on acyl migration in immobilized lipozyme TL-catalyzed transesterification of soybean oil for biodiesel production. *J. Mol. Catal. B: Enzym.* 2005, *37*, 68–71.

Felizardo, P.; Correia, M. J. N.; Raposo, I.; Mendes, J. F.; Berkemeier, R.; Bordado, J. M. Production of Biodiesel from Waste Frying Oils. *Waste Manage.* 2006, *26*, 487–494.

Freedman, B.; Butterfield, R. O.; Pryde, E. H. Transesterification kinetics of soybean oil. *J. Am. Oil Chem. Soc.* 1986, *63*, 1375–80.

Freedman, B.; Pryde E. H.; Mounts T. L. Variables affecting the yields of fatty esters from transesterified vegetable oils. *J. Am. Oil Chem. Soc.* 1984, *61*, 1638–1643.

Fukuda, H.; Kondo, A.; Noda, H. Biodiesel fuel production by transesterification of oils. *J. Biosci. Bioeng.* 2001, *92*, 405–416.

Glišić, S.; Lukić, I.; Skala, D. Biodiesel synthesis athigh pressure and temperature: Analysis of energy consumption on industrial scale. *Bioresour. Technol.* 2009, *100*, 6347–6354.

Goembira, F.; Saka, S. Advanced supercritical Methyl acetate method for biodiesel production from *Pongamia pinnata* oil, *Renewable Energy* 2015, *83*, 1245–1249.

Graboski, M. S.; McCormick, R. L. Combustion of fat and vegetable oil derived fuels in diesel engines. *Prog. Energy Combust. Sci.* 1998, *24*, 125–164.

Gui, M. M.; Lee, K. T. Bhatia, S. Feasibility of edible oil vs. non-edible oil vs. waste edible oil as biodiesel feedstock. *Energy* 2008, *33*, 1646–1653.

Gupta, M. K. Sunflower oil. In *Vegetable Oils in Food Technology*; Gunstone, F. D., Ed.; Blackwell Publishing: Oxford, 2002; 129–157.

Gupta, A. R.; Yadav, S. V.; Rathod, V. K. Enhancement in biodiesel production using waste cooking oil and calcium diglyceroxide as a heterogeneous catalyst in presence of ultrasound. *Fuel* 2015, *158*, 800–806.

Hartman, L. Processo de extracao de oleos vegetais para obtencao de combustiveis. Brazilian Patent PI 8003739, 1980.

Hawash, S.; Kamal, N.; Zaher, F.; Kenawi, O.; El Diwani G. Biodiesel fuel from *Jatropha* oil via non-catalytic supercritical methanol transesterification. *Fuel* 2009, *88*, 579–582.

He, H.; Wang, T.; Zhu, S. Continuous production of biodiesel fuel from vegetable oil using supercritical methanol process. *Fuel* 2007, *86*, 442–447.

Helwani, Z.; Othman, M. R.; Aziz, N.; Fernando, W. J. N.; Kim, J. Technologies for production of biodiesel focusing on green catalytic techniques: A review. *Fuel Process. Technol.* 2009, *90*, 1502–1514.

Ilgen, O. Dolomite as a heterogeneous catalyst for transesterification of canola oil. *Fuel Process. Technol.* 2011, *92*, 452–455.

Ilgen, O.; Akin, A. N. Transesterification of Canola Oil to Biodiesel Using MgO Loaded with KOH as a Heterogeneous Catalyst. *Energy Fuels* 2009, *23*, 1786–1789.

Ilham, Z., Saka, S. Dimethyl carbonate as potential reactant in non-catalytic biodiesel production by supercritical method. *Bioresour. Technol.* 2009, *100*, 1793–1796.

Issariyakul, T.; Dalai, A. K. Biodiesel from vegetable oils. *Renewable Sustainable Energy Rev.* 2014, *31*, 446–471.

Jacobson, K.; Gopinath, R.; Meher, L. C.; Dalai, A. K. Solid acid catalyzed biodiesel production from waste cooking oil. *Appl. Catal., B* 2008, *85*, 86–91.

Jain, S.; Sharma, M. P. Biodiesel production from *Jatropha curcas* oil. *Renewable Sustainable Energy Rev.* 2010, *14*, 3140–3147.

Kansedo, J.; Lee, K. T. Transesterification of palm oil and crude sea mango (*Cerbera odollam*) oil: The active role of simplified sulfated zirconia catalyst. *Biomass Bioenergy* 2012, *40*, 96–10.

Kansedo, J.; Lee, K. T.; Bhatia, S. *Cerbera odollam* (sea mango) oil as a promising non-edible feedstock for biodiesel production. *Fuel* 2009, *88*, 1148–1150.

Karmakar, A.; Karmakar, S.; Mukherjee, S. Properties of various plants and animals feedstocks for biodiesel production. *Bioresour. Technol.* 2010, *101*, 7201–1072.

Karmee, S. K.; Chadha, A. Preparation of biodiesel from crude oil of *Pongamia pinnata. Bioresour. Technol.* 2005, *96*, 1425–1429.

Kaur, M.; Ali, A. Lithium ion impregnated calcium oxide as nano catalystfor the biodiesel production from karanja and jatropha oils. *Renewable Energy* 2011, *36*, 2866–2871.

Kaur, N.; Ali, A.. Kinetics and reusability of Zr/CaO as heterogeneous catalyst for the ethanolysis and methanolysis of *Jatropha crucas* oil. *Fuel Process. Technol.* 2014, *119*, 173–184.

Kesić, Ž.; Lukić, I.; Brkić, D.; Rogan, J.; Zdujić, M.; Liu, H.; Skala, D. Mechanochemical preparation and characterization of CaO·ZnO used as catalyst for biodiesel synthesis. *Appl. Catal., A* 2012, *427–428*, 58–65.

Kesić, Ž.; Lukić, I.; Zdujić, M.; Jovalekić, Č.; Liu, H.; Skala, D. Mechanochemical synthesis of $CaO \cdot ZnO \cdot K_2CO_3$ catalyst: Characterization and activityfor methanolysis of sunflower oil. *Chem. Ind. Chem. Eng. Q* 2015, *21,* 1−12.

Kesić, Ž.; Lukić, I.; Zdujić, M.; Jovalekić, Č.; Veljković, V.; Skala, D. Assessment of $CaTiO_3$, $CaMnO_3$, $CaZrO_3$ and $Ca_2Fe_2O_5$ perovskites as heterogeneous base catalysts for biodiesel synthesis. *Fuel Process. Technol.* 2016, *143*, 162–168.

Knothe, G. Historical perspectives on vegetable oil-based diesel fuels. *Inform* 2001, *12*, 1103–1107.

Knothe, G. Biodiesel and renewable diesel: A comparison. *Prog. Energy Combust. Sci.* 2010, *36*, 364–373.

Knothe, G.; Krahl, J.; Van Gerpen J. The biodiesel handbook. AOCS Press: Champaign, IL, USA, 2005.

Kondamudi, N.; Mohapatra, S. K.; Misra, M. Quintinite as a bifunctional heterogeneous catalyst for biodiesel synthesis. *Appl. Catal., A* 2011, *393*, 36–43.

Kouzu, M.; Kasuno, T.; Tajika, M.; Sugimoto, Y.; Yamanaka, S.; Hidaka, J. Calcium oxide as a solid base catalyst for transesterification of soybean oil and its application to biodiesel production. *Fuel* 2008, *87*, 2798–2806.

Körbitz, W. Biodiesel production in Europe and North America, an encouraging prospect. *Renewable Energy* 1999, *16*, 1078–1083.

Kulkarni, M.G.; Dalai, A. K. Waste cooking oils an economical source for biodiesel: A review. *Ind. Eng. Chem. Res.* 2006, *45*, 2901–2913.

Kumar, A.; Sharma, S. Potential non-edible oil resources as biodiesel feedstock: an Indian perspective. *Renewable Sustainable Energy Rev.* 2011, *15*, 1791–800.

Kumari, V.; Shah, S.; Gupta, M. N. Preparation of biodiesel by lipase-catalyzed transesterification of high free fatty acid containing oil from *Madhuca indica. Energy Fuels* 2007, *21*, 368–372.

Kusdiana, D.; Saka, S. Biodiesel fuel from rapeseed oil as prepared in supercritical methanol. *J. Chem. Eng. Jpn.* 2001, *34*, 383–387.

Kusdiana, D.; Saka, S. Effects of water on biodiesel fuel production by supercritical methanol treatment. *Bioresour. Technol.* 2004a, *91*, 289–295.

Kusdiana, D.; Saka, S. Two-step preparation for catalyst-free biodiesel fuel production. *Appl. Biochem. Biotechnol.* 2004b, *115*, 781–791.

Lajara, J. R.; Diaz, U.; Quidiello, R. D. Definite influence of location and climatic conditions on the fatty acid composition of sunflower seed oil. *J. Am. Oil Chem. Soc.* 1990, *67*, 618–623.

Lam, M. K.; Lee, K. T. Accelerating transesterification reaction with biodiesel as co-solvent: A case study for solid acid sulfated tin oxide catalyst. *Fuel* 2010, *89*, 3866–3870.

Lee, S.; Posarac, D.; Ellis, N. An experimental investigation of biodiesel synthesisfrom waste canola oil using supercritical methanol. *Fuel* 2012, *91*, 229–37.

Lee, J. S.; Saka, S. Biodiesel production by heterogeneous catalysts and supercritical technologies. *Bioresour. Technol.* 2010, *101*, 7191–7200.

Leung, D. Y. C.; Guo, Y. Transesterification of neat and used frying oil: Optimization for biodiesel production. *Fuel Process. Technol.* 2006, *87*, 883–890.

Leung, D. Y. C.; Wu, X.; Leung, M. K. H. A review on biodiesel production using catalyzed transesterification. *Appl. Energy* 2010, *87*, 1083–1095.

Li, L.; Du, W.; Liu, D.; Wang, L.; Li Z. Lipase-catalyzed transesterification of rapeseed oils for biodiesel production with a novel organic solvent as the reaction medium. *J. Mol. Catal. B: Enzym.* 2006, *43 (1)*, 58–62.

Li, N.-W.; Zong, M.-H.; Wu, H. Highly efficient transformation of waste oil to biodiesel by immobilized lipase from *Penicillium expansum. Process Biochem.* 2009, *44*, 685–688.

Limmanee, S.; Naree, T.; Bunyakiat, K.; Ngamcharussrivichai, C. Mixed oxides of Ca, Mg and Zn as heterogeneous base catalysts for the synthesis of palm kernel oil methyl esters. *Chem. Eng. J.* 2013, *225*, 616–624.

Lin, S. W.; Palm oil. In *Vegetable Oils in Food Technology*; Gunstone, F. D., Ed.; Blackwell Publishing: Oxford, 2002; 59–98.

Liu, L.; Wen, Z.; Cui, G. Preparation of Ca/Zr mixed oxide catalysts through a birch-templating route for the synthesis of biodiesel via transesterification. *Fuel* 2015, *158*, 176–182.

López Granados, M.; Martín Alonso, D.; Sádaba, I.; Mariscal, R.; Ocón, P. Leaching and homogeneous contribution in liquid phase reaction catalyzed by solids: The case of triglycerides methanolysis using CaO. *Appl. Catal., B* 2009, *89*, 265–272.

López Granados, M.; Zafra Poves, M. D.; Alonso, D. M.; Mariscal, R.; Galisteo, F. C.; Moreno-Tost, R.; Santamaría, J.; Fierro, J. L. G. Biodiesel from sunflower oil by using activated calcium oxide. *Appl. Catal., B* 2007, *73*, 327–336.

Lukić, I.; Kesić, Ž.; Maksimović S.; Zdujić, M.; Krstić, J.; Skala, D. Kinetics of heterogeneous methanolysis of sunflower oil with Cao·Zno catalyst: influence of different hydrodynamic conditions. *Chem. Ind. Chem. Eng. Q.* 2014, *20 (3)*, 425–439.

Lukić, I.; Kesić, Ž.; Skala, D. Kinetics of Heterogeneous Biodiesel Synthesis Using Supported ZnO as Catalyst. *Chem. Eng. Technol.* 2014, *37*, 1–7.

Lukić, I.; Kesić, Ž.; Zdujić, M.; Skala, D. Calcium diglyceroxide synthesized by mechanochemical treatment, its characterization and application as catalyst for fatty acid methyl esters production. *Fuel* 2016, *165*, 159–165.

Lukić, I.; Krstić, J.; Jovanović, D.; Skala, D. Alumina/silica supported K_2CO_3 as a catalyst for biodiesel synthesis from sunflower oil. *Bioresour. Technol.* 2009, *100*, 4690–4696.

Lukić, I.; Krstić, J.; Glišić, S.; Jovanović, D.; Skala, D. Biodiesel synthesis using K_2CO_3/Al-O-Si aerogel catalyst. *J. Serb. Chem. Soc.* 2010, *75*, 789–801.

Ma, F.; Hanna, M. A. Biodiesel production: a review. *Bioresour. Technol.* 1999, *70*, 1–15.

Maddikeri, G. L.; Pandit, A. B.; Gogate, P. R. Intensification approaches for biodiesel synthesis from waste cooking oil: A review. *Ind. Eng. Chem. Res.* 2012, *51*, 14610−14628.

Madras, G.; Kolluru, C.; Kumar R. Synthesis of biodiesel in supercritical fluids. *Fuel* 2004, *83*, 2029–2033.

Meher, L. C.; Dharmagadda Vidya, S. S.; Naik, S. N. Optimization of alkali-catalyzed transesterification of *Pongamia pinnata* oil for production of biodiesel. *Bioresour. Technol.* 2006, *97*, 1392–1397.

Mekhilefa, S.; Siga, S.; Saidur, R. A review on palm oil biodiesel as a source of renewable fuel. *Renewable Sustainable Energy Rev.* 2011, *15*, 1937–1949.

Miladinović, M.; Lukić, I.; Stamenković, O.; Veljković, V.; Skala, D. Heterogeneous base-catalyzed methanolysis of vegetable oils: state of art. *Chem. Industry* 2010, *64 (2)*, 63–80 (in Serbian language).

Milazzo, M. F.; Spina, F.; Cavallaro, S.; Bart, J. C. J. Sustainable soy biodiesel. *Renewable Sustainable Energy Rev.* 2013b, *27*, 806–852.

Milazzo, M. F.; Spina, F.; Vinci, A.; Espro, C.; Bart, J. C. J. Brassica biodiesels: Past, present and future. *Renewable Sustainable Energy Rev.* 2013a, *18*, 350–389.

Modi, M. K.; Reddy, J. R.; Rao B. V.; Prasad, R. B. Lipase-mediated transformation of vegetable oils into biodiesel using propan-2-ol as acyl acceptor. *Biotechnol. Lett.* 2006, *28(9)*, 637-640.

Modi, M. K.; Reddy, J. R. C.; Rao, B. V. S. K.; Prasad, R. B. N. Lipase-mediated conversion of vegetable oils into biodiesel using ethyl acetate as acyl acceptor. *Bioresour. Technol.* 2007, *98*, 1260–1264.

Mofijur, M.; Atabani, A. E.; Masjuki, H. H.; Kalam, M. A.; Masum, B. M. A study on the effects of promising edible and non-edible biodiesel feedstocks, on engine performance and emissions production: A comparative evaluation. *Renewable Sustainable Energy Rev.* 2013, *23*, 391–404.

Moradi, G.; Mohadesi, M.; Hojabri, Z. Biodiesel production by CaO/SiO_2 catalyst synthesizedby the sol–gel process. *React. Kinet., Mech. Catal.* 2014, *113 (1)*, 169–186.

Niu, S. L.; Huo, M. J.; Lu, C. M.; Liu, M. Q.; Li, H. An investigation on the catalytic capacity of dolomite in transesterification and the calculation of kinetic parameters. *Bioresour. Technol.* 2014, *158*, 74–80.

Noiroj, K.; Intarapong, P.; Luengnaruemitchai, A.; Jai-In, S. A comparative study of KOH/Al_2O_3 and KOH/NaY catalysts for biodiesel production via transesterification from palm oil. *Renewable Energy* 2009, *34*, 1145–1150.

Noureddini, H.; Zhu, D. Kinetics of transesterificatio of soybean oil. *J. Am. Oil Chem. Soc.* 1997, *74*, 1457–63.

O'Brien R. Cottonseed oil. In *Vegetable Oils in Food Technology*; Gunstone, F. D., Ed.; Blackwell Publishing: Oxford, 2002; 203–231.

Ognjanovic, N.; Bezbradica, D.; Knezevic-Jugovic, Z. Enzymatic conversion of sunflower oil to biodiesel in a solvent-free system: Process optimization and the immobilized system stability. *Bioresour. Technol.* 2009, *100*, 5146–5154.

Olutoye, M. A.; Hameed, B. H. Synthesis of fatty acid methyl ester from crude jatropha (*Jatropha curcas Linnaeus*) oil using aluminium oxide modified Mg–Zn heterogeneous catalyst. *Bioresour. Technol.* 2011, *102*, 6392–6398.

Olutoye, M. A.; Hameed, B. H. Production of biodiesel fuel by transesterification of different vegetable oils with methanol using Al_2O_3 modified MgZnO catalyst. *Bioresour. Technol.* 2013, *132*, 103–108.

Onoji, S. E.; Iyuke, S. E.; Igbafe, A. I.; Nkazi, D. B. Rubber seed oil: A potential renewable source of biodiesel for sustainable development in sub-Saharan Africa. *Energy Convers. Manage.* 2016, *110*, 125–134.

Pantzaris, T. P.; Basiron, Y. The lauric (coconut and palmkernel) oils. In *Vegetable Oils in Food Technology*; Gunstone, F. D., Ed.; Blackwell Publishing: Oxford, 2002; 157–203.

Patil, P. D.; Gude, V. G.; Deng, S. Biodiesel pProduction from *Jatropha Curcas*, waste cooking, and *Camelina sativa* oils. Ind. Eng. Chem. Res. 2009, *48*, 10850–10856.

Przybylski, R.; Mag, T. Canola/rapeseed oil, In *Vegetable Oils in Food Technology*; Gunstone, F. D., Ed.; Blackwell Publishing: Oxford, 2002; 98–128.

Puna, J. F.; Correia, M. J. N.; Dias, A. P. S.; Gomes, J.; Bordado, J. Biodiesel production from waste frying oils over lime catalysts. *React. Kinet., Mech. Catal.* 2013, *109*, 405–415.

Raita, M.; Arnthong, J.; Champreda, V.; Laosiripojana N. Modification of magnetic nanoparticle lipase designs for biodiesel production from palm oil. *Fuel Process. Technol.* 2015a, *134*, 189–197.

Raita, M.; Laosiripojana, N.; Champreda, V. Biocatalytic methanolysis activities of cross-linked protein-coated microcrystalline lipase toward esterification/transesterification of relevant palm products. *Enzyme Microb. Technol.* 2015b, *70*, 28–34.

Ramos, M. J.; Fernández, C. M.; Casas, A.; Rodríguez, L.; Pérez, A. Influence offatty acid composition of raw materials on biodiesel properties. *Bioresour. Technol.* 2009, *100*, 261–268.

Rashid, U.; Anwar, F. Production of biodiesel through optimizedalkaline-catalyzed transesterification of rapeseed oil. *Fuel* 2008, *87*, 265–73.

Rathore, V.; Madras, G. Synthesis of biodiesel from edible and non-edible oils in supercritical alcohols and enzymatic synthesis in supercritical carbon dioxide. *Fuel* 2007, *86*, 2650–2659.

Reddy, C. R. V.; Oshel, R.; Verkade, J. G. Room-temperature conversion of soybean oil and poultry fat to biodiesel catalyzed by nanocrystalline calcium oxides. *Energy Fuels* 2006, *20*, 1310–1314.

Refaat, A. A. Biodiesel production using solid metal oxide catalysts. *Int. J. Environ. Sci. Technol.* 2011, *8*, 203–221.

Refaat, A. A.; Attia, N. K.; Sibak, H. A.; Elsheltawy, S. T.; Diwani G. I. Production optimization andquality assessment of biodiesel form waste vegetable oil. *Int. J. Environ. Sci. Technol.* 2008, *5(1)*, 75–82.

Reinoso, D. M.; Damiani, D. E.; Tonetto, G. M. Zinc glycerolate as a novel heterogeneous catalyst for the synthesis of fatty acid methyl esters. *Appl. Catal., B* 2014, *144*, 308–316.

Ribeiro, L. M. O.; da S. Santos, B. C.; Almeida, R. M. R. G. Studies on reaction parameters influence on ethanolic production of coconut oil biodiesel using immobilized lipase as a catalyst. *Biomass Bioenergy* 2012, *47*, 498–503.

Royon, D.; Daz, M.; Ellenrieder, G.; Locatelli, S. Enzymatic production of biodiesel from cotton seed oil using t-butanol as a solvent. *Bioresour. Technol.* 2007, *98*, 648–653.

Saka, S.; Isayama, Y. A new process for catalyst-free production of biodieselusing supercritical methyl acetate. *Fuel* 2009, *88*, 1307–1313.

Saka, S.; Kusdiana, D. Biodiesel fuel from rapeseed oil as prepaired in suprcritical methanol. *Fuel* 2001, *80*, 225–231.

Samniang, A.; Tipachan, C.; Kajorncheappun-ngam, S. Comparison of biodiesel production from crude *Jatropha oil* and *Krating oil* by supercritical methanol transesterification. *Renewable Energy* 2014, *68*, 351–355.

Saravanan, N.; Puhan, S.; Nagarajan, G.; Vedaraman, N. An experimental comparison of transesterification process with different alcohols using acid catalysts. *Biomass Bioenergy* 2010, *34*, 999–1005.

Sánchez-Cantú, M.; Pérez-Díaz, L. M.; Pala-Rosas, I.; Cadena-Torres, E.; Juárez-Amador, L.; Rubio-Rosas, E.; Rodríguez-Acosta, M.; Valente, J. S. Hydrated lime as an effective heterogeneous catalyst for the transesterification of castor oil and methanol. *Fuel* 2013, *110*, 54–62.

Schwab, A. W.; Bagby, M.O.; Freedman, B. Preparation and properties of diesel fuels from vegetable oils. *Fuel* 1987, *66*, 1372–1378.

Shah, S.; Gpta, M. N. Lipase catalyzed preparation of biodiesel from *Jatropha* oil in a solvent free system. *Process Biochem.*2007, *42*, 409–414.

Sharma, Y. C.; Singh B. A hybrid feedstock for a very efficient preparation of biodiesel. *Fuel Process. Technol.* 2010, *91*, 1267–1273.

Singh, B.; Guldhe, A.; Rawat, I.; Bux, F. Towards a sustainable approach for development of biodiesel from plant and microalgae. *Renewable Sustainable Energy Rev.* 2014, *29*, 216–245.

Skala, D.; Glišić, S.; Lukić, I.; Orlović, A. Biodiesel II: A new concept of biodiesel production-transesterification with supercritical methanol. *Chem. Industry* 2004, *58 (4)*, 176–185 (in Serbian language).

Soares Dias, A. P.; Puna, J.; Correia, M. J. N.; Nogueira, I.; Gomes, J.; Bordado, J. Effect of the oil acidity on the methanolysis performances of lime catalyst biodiesel from waste frying oils (WFO). *Fuel Process. Technol.* 2013, *116*, 94–100.

Song, R.; Tong, D.; Tang, J.; Hu, C. Effect of composition on the structure and catalytic properties of KF/MgLa solid base catalysts for biodiesel synthesis via transesterification of cottonseed oil. *Energy Fuels* 2011, *25*, 2679–2686.

Soumanou, M. M.; Bornscheuer, U. T. Improvement in lipase-catalyzed synthesis of fatty acid methyl esters from sunflower oil. *Enzyme Microb. Technol.* 2003, *33*, 97–103.

Sree, R.; Babu, N. S.; Prasad, P. S. S.; Lingaiah, N. Transesterification of edible and non-edible oils over basic solid Mg/Zr catalysts. *Fuel Process. Technol.* 2009, *90*, 152–157.

Srilatha, K; Prabhavathi Devi, B. L. A.; Lingaiah, N.; Prasad, R. B. N.; Sai Prasad, P. S. Biodiesel production from used cooking oil by two-step heterogeneous catalyzed process. *Bioresour. Technol.* 2012, *119*, 306–311.

Srivastava, A.; Prasad, R. Triglycerides-based diesel fuels. *Renewable Sustainable Energy Rev.* 2000, *4*, 111–133.

Stojković, I. J.; Stamenković, O. S.; Povrenović, D. S.; Veljković, V. B. Purification technologies for crude biodiesel obtained by alkali-catalyzed transesterification. *Renewable Sustainable Energy Rev.* 2014, *32*, 1–15.

Suppes, G. J.; Dasari, M. A.; Doskocil, E. J.; Mankidy, P. J.; Goff, M. J. Transesterification of soybean oil with zeolite and metal catalysts. *Appl. Catal., A* 2004, *254*, 213–223.

Talebian-Kiakalaieh, A.; Amin, N. A. S.; Mazaheri, H. A review on novel processes of biodiesel production from waste cooking oil. *Appl. Energy* 2013, *104*, 683–710.

Tamalampudi, T.; Talukder, M. R.; Hama, S.; Numata, T.; Kondo, A.; Fukuda, H. Enzymatic production of biodiesel from Jatropha oil: A comparative study of immobilized-whole cell and commercial lipases as a biocatalyst. *Biochem. Eng. J.*2008, *39*, 185–189.

Tan, K. T.; Lee, K. T.; Mohamed A. R. Potential of waste palm cooking oil for catalystfree biodiesel production. *Energy* 2011, *36*, 2085–2088.

Tanaka, Y.; Ichikawa, C.; Okabe, A.; Ando, S. Verfahren zur Herstellung niederer fettsaureaalkylester. German Patent EP 1 126 011 A2, 2001.

Tang, Z.; Wang, L.; Yang, J. Transesterification of the crude *Jatropha curcas L.* oil catalyzed by micro-NaOH in supercritical and subcritical methanol. *Eur. J. Lipid Sci. Technol.* 2007, 109, 585–590.

Tapanes, N. C. O.; Aranda, D. A. G.; Carneiro, J. W. D. M.; Antunes, O. A. C. Transesterification of *Jatropha curcas* oil glycerides: theoretical and experimental studies of biodiesel reaction. *Fuel* 2008, *87*, 2286–2295.

Taufiq-Yap, Y. H.; Lee, H. V.; Hussein, M. Z.; Yunus, R. Calcium-based mixed oxide catalysts for methanolysis of *Jatropha curcas* oil to biodiesel. *Biomass Bioenergy* 2011, *35*, 827–834.

United States Department of Agriculture. http://apps.fas.usda.gov/psdonline/circulars/oilseeds.pdf (Accessed Jan 5, 2016).

Usta, N. Use of tobacco seed oil methyl ester in a turbocharged indirect injection diesel engine. *Biomass Bioenergy* 2005, *28*, 77–86.

Varma, M. N.; Madras, G. Synthesis of Biodiesel from Castor Oil and Linseed Oil in Supercritical Fluids. *Ind. Eng. Chem. Res.* 2007, *46*, 1–6.

Vyas, A. P.; Subrahmanyama, N.; Patel, P. A. Production of biodiesel through transesterification of *Jatropha* oil using KNO_3/Al_2O_3 solid catalyst. *Fuel* 2009, *88*, 625–628.

Veljković, V. B.; Lakićević, S. H.; Stamenković, O. S.; Todorović, Z. B.; Lazić, M. L. Biodiesel production from tobacco (*Nicotiana tabacum L.*) seed oil with a high content of free fatty acids. *Fuel* 2006, *85*, 2671–2675.

Veljković, V. B.; Stamenković, O. S.; Todorović, Z. B.; Lazić, M. L.; Skala, D. U. Kinetics of sunflower oil methanolysis catalyzed by calcium oxide. Fuel 2009, *88*, 1554–1562.

Vicente, G.; Martinez, M.; Aracil, J. Integrated biodiesel production: a comparison of different homogeneous catalysts systems. *Bioresour. Technol.* 2004, *92*, 297–305.

Viola, E.; Blasi, A.; Valerio, V.; Guidi, I.; Zimbardi, F.; Braccio, G.; Giordano, G. Biodiesel from fried vegetable oils via transesterification by heterogeneous catalysis. *Catal. Today* 2012, *179*, 185–190.

Wang, T. Soybean oil. In *Vegetable Oils in Food Technology*; Gunstone, F. D., Ed.; Blackwell Publishing: Oxford, 2002; 18–58.

Wang, L.; Yang, J. Transesterification of soybean oil with nano-MgO or not in supercritical and subcritical methanol. *Fuel* 2007, *86*, 328–333.

Wang, Y.; Ou, S.; Liu, P.; Zhang, Z.; Preparation of biodiesel from waste cooking oil via two-step catalyzed process. *Energy Convers. Manage.* 2007, *48*, 184–188.

Wang, Y.; Ou, S.; Liu, P.; Xue, F.; Tang, S. Comparison of two different processes to synthesize biodiesel by waste cooking oil. *J. Mol. Catal. A: Chem.* 2006, *252*, 107–112.

Warabi, Y.; Kusdiana, D.; Saka, S. Reactivity of triglycerides and fatty acids of rapeseed oil in supercritical alcohols. *Bioresour. Technol.* 2004, *91*, 283–287.

Xie, W.; Peng, H.; Chen, L. Calcined Mg–Al hydrotalcites as solid base catalysts for methanolysis ofsoybean oil. *J. Mol. Catal. A: Chem.* 2006, *246*, 24–32.

Xie, W.; Wang, J. Immobilized lipase on magnetic chitosan microspheres for transesterification of soybean oil. *Biomass Bioenergy* 2012, *36*, 373–380.

Yaakob, Z.; Mohammad, M.; Alherbawi, M.; Alam, Z.; Sopian, K. Overview of the production of biodiesel from Waste cooking oil. *Renewable Sustainable Energy Rev.* 2013, *18*, 184–193.

Yan, J.; Yan, Y.; Liu, S.; Hu, J.; Wang, G. Preparation of cross-linked lipase-coated micro-crystals for biodiesel production from waste cooking oil. *Bioresour. Technol.* 2011, *102*, 4755–4758.

Yan, S.; Kim, M.; Salley, S. O.; Ng, K. Y. S. Oil transesterification over calcium oxides modified with lanthanum, *Appl. Catal., A* 2009, *360*, 163–170.

Yan, S.; Lu, H.; Liang, B. Supported CaO catalysts used in the transesterification of rapeseed oil for the purpose of biodiesel production. *Energy Fuels* 2008, *22*, 646–651.

You, O.; Yin, X.; Zhao, Y.; Zhang, Y. Biodiesel production from jatropha oil catalyzed by immobilized *Burkholderiacepacia* lipase on modified attapulgite. *Bioresour. Technol.*2013, *148*, 202–207.

Zabeti, M.; Daud W. M. A. W.; Aroua M. K. Optimization of the activity of CaO/Al_2O_3 catalyst for biodiesel production using response surface methodology. *Appl. Catal., A* 2009, *366*, 154–159.

Zhang, Y.; Dube, M. A.; McLean, D. D.; Kates, M. Biodiesel production from waste cooking oil: 1. Process design and technological assessment. *Bioresour. Technol.* 2003, *89*, 1–16.

Zhang, J.; Chen, S.; Yang, R.; Yan, Y. Biodiesel production from vegetable oil using heterogenous acid and alkali catalyst. *Fuel* 2010, *89*, 2939–2944.

Zhao, L.; Qiu, Z.; Stagg-Williams, S. M. Transesterification of canola oil catalyzed by nanopowder calcium oxide. *Fuel Process. Technol.* 2013, *114*, 154–162.

Zheljazkov, V. D.; Vick, B. A.; Baldwin, B. S.; Buehring, N.; Coker, C.; Astatkie, T.; Johnson, B. Oil productivity and composition of sunflower as a function of hybrid and planting date. *Ind. Crops Prod.* 2011, *33*, 537–543.

Zhu, H.; Wu, Z.; Chen, Y.; Zhang, P.; Duan, S.; Liu, X.; Mao, Z. Preparation of biodiesel catalyzed by solid super base of calcium oxide and its refining process. *Chin. J. Catal.* 2006, *27*, 391–396.

Zięba, A.; Pacuza, A; Drelinkiewicz, A. Transesterification of triglycerides with methanol catalyzed by heterogeneous zinc hydroxy nitrate catalyst. Evaluation of variables affecting the activity and stability of catalyst. *Energy Fuels* 2010, *24*, 634–645.

In: Vegetable Oil
Editor: Brittany Holt

ISBN: 978-1-63485-128-2

Chapter 5

OZONIZED VEGETABLE OILS: PRODUCTION, CHEMICAL CHARACTERIZATION AND THERAPEUTIC POTENTIAL

Nathália Rodrigues de Almeida[1,*], Adilson Beatriz[1], Eduardo José de Arruda[2], Dênis Pires de Lima[1], Lincoln Carlos Silva de Oliveira[1] and Ana Camila Micheletti[1]

[1]Universidade Federal do Mato Grosso do Sul, UFMS, Brazil
[2]Universidade Federal da Grande Dourados, UFGD, Brazil

ABSTRACT

Ozonated vegetable oils have demonstrated promising results for clinical application, and they have been the focus of great pharmaceutical interest to treat dermatological disorders, such as infections of skin ulcers and chronic wounds. There are reports of these products as effective to heal refractory wounds, where conventional treatments and available medications prove ineffective. In fact, in some European countries, such as Germany, they can be obtained on prescription from pharmacies. Countries such as Cuba have developed commercial ozonated oils, and they have been successfully tested to treat many diseases. Cuba is one of the pioneers in the implementation of this therapy in Public Health Services for over 22 years. Ozone reacts with the double bonds of unsaturated fatty acids of vegetable oils, providing stable ozonation products, mainly ozonides, hydroperoxides and polyperoxides (depending on reaction conditions) with therapeutic potential. Several studies have demonstrated their antimicrobial and antifungal activity, as well as their role as wound healing modulators, showing no cytotoxicity when tested against NIH/3T3 murine fibroblast cells. Simple analytical techniques such as peroxide value, iodine value and viscosity determination have been extensively used for characterization of products, together with spectroscopic techniques of NMR (^{1}H and ^{13}C) and infrared, chromatography and thermal analysis (TG/DTG - DSC). This chapter aims to highlight recent contributions to the production, characterization and biological activities of ozonated vegetable oils.

[*] Corresponding author: Email: nathrodrigues@live.com.

Keywords: ozonized vegetable oil, ozonides, antimicrobial activity

INTRODUCTION

Oils or vegetable fats are lipids (glycerol esters) extracted from seeds, nuts and native fruits or short crop cycle or long/perennial. They can be extracted by mechanical (molding) or by chemical solvent extraction or a combination of these methods [1].

The oils are products rich in monounsaturated fatty chains and poly-unsaturated acids, antioxidants, fat-soluble vitamins and other components in minor proportions. Vegetable oils have low allergenic capacity, are nontoxic and used as fuel, food, functional product due to the intrinsic therapeutic properties or structural, cosmetic changes, industrial and pharmaceutical uses [2].

The cutaneous application of vegetable oil as a therapeutic practice and dietary coadjuvant has been described mainly in adult patients at risk for essential fatty acid deficiency [3]. Vegetable oils and volatile oils are able to increase the potential penetration of different drugs. Formulations with different vegetable oils for medicinal purposes facilitates transdermal permeation, mainly due to entrainment efficiency and bioavailability [4]. The concentration of vegetable oils in the formulations increases permeation and this is attributed to the fatty acids content as confirmed by studies on release kinetics and transdermal permeation [4, 5].

Modification of unsaturated fatty acids by ozonation at a different time is an option to carry out the synthesis of new drugs and/or preparation of different pharmaceutical forms. The ozonation fatty acids and chemical transformation allow preparation of a single product carrier which is bioactive and at the same time for multiple purposes. Unsaturated vegetable oils can be ionized or produced as oleogel for direct filling. The oleogel has spreadability, adhesiveness and melting temperature close to body temperature. Assays with gram-positive and gram-negative resistant bacteria strains and tumor cells showed a biological activity of vegetable oils and their applicability for wound healing and treatment of chronic wounds. This chapter discusses the synthesis, mechanisms, characterization and biological activity of ozonated vegetable oils for healing chronic wounds.

The chemical synthesis for obtaining ozonized vegetable oils using unsaturated oils, bubbling a gas mixture (O_2/O_3) of a porous ceramic for different periods of time in a reactor. The preparation of oleogel is by its deposition on a water bed (up to 10%) in contact with the porous ceramic to distribute the gas and O_3 reaction with unsaturated vegetable oil fatty acids to form ozonides and lipoperoxides (aldehydes, ketones, peroxides) [6]. Ozone reaction with unsaturated constituents of vegetable oils occurs exclusively with the carbon-carbon double bonds of unsaturated acids and ozonides to produce lipid peroxides. Various oxygen compounds are formed depending on the ozonation time: aldehydes, diperoxides and polyperoxides; which are partly responsible for part of the biological activity of ozonated oils [6, 7].

Despite simplicity of the reaction which normally uses direct ozone bubbling in the vegetable oil, the quality and process reproducibility for obtaining oils patterns depend on the quality and purity from vegetable oils and other process parameters, such as: (i) the type of ozone generators; (ii) the conditions of ozonation, reactor, ozonation time, reactor material and process scale iii) the presence of water or catalysts; (iv) the efficiency of the ozonator, flow and concentration of O_3 carrier gas/diluent and yet the purity of oxygen used in addition to other parameters [8].

The characterization of the species produced in the ozonation needs to be performed, as well as details of the reaction kinetics at different process times. The physicochemical properties of the products obtained in different ozonation times should also be assigned. Accordingly, the quality of ozonized products must be checked by various spectroscopic and chromatographic techniques and/or classical analytical techniques for the complete characterization and accurate quantification of ozonized products. The most widely used techniques in the ozonation are Fourier Transformed Infrared (FT-IR), $^{1}H/^{13}C$-NMR and analytical methods for the determination of peroxide levels, acidity, viscosity and iodine value [9].

The ozone and ozonized product from the ozone reaction with fatty acids and other substrates have intense biological activity (bactericidal, virucidal, etc.). Ozone disinfectant properties are well revealed in the use of ozonated vegetable oil. Ozonated oil was found to have antiseptic activity. These products stimulate the immune system and act on the healing and tissue repair. The biological activity of the product stability and performance in different formulations allow its use in pre-clinical and clinical testing [10]. Unsaturated fatty acids are essential components of vegetable oils and cellular membranes, therefore, the effects of their ozonation have been widely studied. In the literature, many studies are available using different vegetable oils, especially in topical applications (external use), although there are reports of the oral administration. In these studies, the majority of vegetable oils used for these different purposes are the olive and sunflower oils. However, different vegetable oils or essential oils can be used for producing ozonized products for different cosmetic or therapeutic applications. Ozonized commercial oils are available for a variety of applications. Unsatured oils are best known and available for nutritional purposes [11]. The composition of unsaturated fatty acid are chemical intermediates for ozonation and production of several products for treating of skin (acne, herpes, wrinkle, psoriasis, cracking and peeling), treatment and healing of chronic wounds and other applications. Some examples of such oils are the sunflower, thyme, sesame, soybean, coconut, olive, hemp seed, grape seed, jojoba, sweet almond, avocado, flax seed, and others.

Currently, to treat skin diseases, there are a variety of anti-infective agents available. Some products of these commercial products for topical use have become inefficient due to the emergence of resistance of pathogens. The ozone has strong biological activity, eliminate pathogens from the release of oxygen species, while active fibroblast proliferation, induces the reconstruction of the intercellular matrix, keratinoblasts proliferation, and wound healing. Accordingly, new products have been proposed for the synthesis and its proper use to meet the demands of the pharmaceutical industry.

In this scenario, the use of ozone and vegetable oils allow the production of new assets that are both bioactive and vehicle with excellent antibacterial properties, fungicides and, with a healing property. The ozone is an unstable molecule with intense biological activity by its decomposition into singlet oxygen. This species is highly reactive to pathogens (viruses, bacteria, protozoa and other microorganisms) and can degrade natural or synthetic compounds and their metabolites residues. The produced ozonated derivatives of unsaturated oily substrates allow the production of a multiplicity of commercial products of wide application. The control difficulties of the oxidative process in the production and topical application, stability, and adequacy of the application form have been carried out by new techniques, pharmaceutical forms and strategies in the ozonation of vegetable oils, in addition to increasing the stability and conservation of characteristics of the obtained products [9].

Numerous diseases can lead to failure of organs, mucous membranes and arteries. This problem has been exacerbated with aging and many skin lesions are subject, ulcers and generalized infections with high socioeconomic cost. In this respect, Zanardi et al. (2013) [12] shows that ozonized oils are a low-cost option compared to conventional antibiotics for topical application to the reduction of the infection, and wound healing. The intense bactericidal and immunostimulant effect of these products accelerates healing. The study also shows that the removal of debris and exudates is a necessary increase in the bactericidal effect of the ozonated condition products [12].

The effect of the ozonization in olive oil, soybean oil, oleic-, linoleic- and linolenic acid was performed by Sadowskaet et al. [13]. The products obtained from different time ozonation were analyzed by different techniques: ^{1}H NMR [14], ^{13}C NMR. The amounts of peroxide and acid contents, viscosity and molecular weight were determined in pure and ozonized oils. Results showed that the chains of fatty acids showed a gradual decrease in the unsaturation level with increasing time of ozonation. The reaction products were identified by Criegee mechanism [15]. The produced ozonides and disappearance of the unsaturations were sequential, with increasing values of the indices peroxide and acid to the oils, the highest increases were observed for the soybean oil. The long time ozonation produced a number of products with different molecular weights were identified as oligomeric ozonides and cross-ozonides respectively [13].

A recent study by Moureu et al. [16] reported that ozonation conditions change the composition of ozonated oils, opening new perspectives for the application of ozonated oil. The study showed that the consumption of double bonds is the same carrying out the ozonation conditions with or without water. However, the emphasis has been pointed out that water increases the formation of peroxides and acid species. This result indicates that the bioactivity of the product and the bactericidal activity is better for ozonized oils in the presence of water and this biological activity is specifically related to the existence of peroxides species [16].

Production and Chemical Characterization of Ozonized Vegetable Oils

Production of Ozonized Vegetable Oils

The ozone gas used to the reaction is generated from extra-dry oxygen (99.9%). Vegetable oils are ozonated in a semi-batch reactor or Drechsel bottle with a sintered filter at the bottom using ozone generator with the ozone concentration and gas flow defined [17, 18]. Guerra-Blanco describes the use of ozone analyzer connected to a computer, capable of detecting ozone in gas phase outlet in the reactor. The ozone monitoring is used to measure the ozone consumption and degree of ozonation [17]. Despite there are highly sophisticated ozonolysis equipments, it is possible to build an ozone generator using parts easily obtained on the market and its performance is perfectly suitable for common laboratory work [19].

The efficiency of ozone reaction with ethyl oleate has been monitored by iodometry. The unreacted ozone emerging from the bubble reactor was trapped in a KI and determined each hour by titration with $Na_2S_2O_3$ (sodium thiosulfate) 0.1 M. Ozone reaction with ethyl oleate is nearly quantitative for several hours, and the unreacted O_3 emerges from the bubble reactor when the molar ratio $O_3/C{=}C \approx 1$ is reached [18].

The ozonation of different vegetable oils, such as sunflower, olive, grape, flaxseed, baru, coconut, sesame and canola oil under different conditions have been studied. However, the comparison between the results remains difficult due to the large amount of parameters influencing the reaction, such as ozone flow rate and concentration, and use of solvent. Table 1 shows some reactions conditions.

Ozone reacts with carbon–carbon double bonds unsaturated fatty acids (free and as esters in triglycerides), according to the mechanism described by Criegee [15] to form different oxygenated species such as ozonides, hydroperoxides, polymeric peroxides and aldehydes, depending on reaction condictions [16, 18].

Scheme 1 shows the mechanism of ozonolysis reaction described by Criegee in 1975 [15]. The first product of the reaction is called malozonide (1), which is very unstable and decomposes to give a zwitterion (2) and a carbonyl compound (aldehyde or ketone, depending of starting material) (3). In the absence of polar solvent, the zwitterion must react either with itself or with the carbonyl compound. Reaction with the carbonyl compound (3) to form a monomeric ozonide (4) as the major product and ozonides polymeric as minor products. The zwitterion (2) generally dimerizes to form 5 or polymerizes when 3 is a ketone, less susceptible to nucleophilic attack [11]. In the presence of protic solvents, such as water or alcohol, the zwitterion interact with the solvent to give hydroperoxides (4) in high yield, since the concentration of the solvent far exceeds that of any other substances with which the zwitterion may react.

Spectroscopy

Several techniques have been used to characterize ozonized vegetable oils. Spectroscopic methods are adequate because they make it possible to detect chemical changes that occur from the ozone reaction. Fourier transform infrared (FT-IR) and nuclear magnetic resonance (^{1}H and ^{13}C NMR) are used to study these reactions where the signals corresponding to reactant and product groups are well identified.

In FT-IR spectra of oils before ozonolysis reaction is observed characteristic bands of C=C double bond stretching (1651 - 1654 cm $^{-1}$) and C=C–H stretching (3009 cm^{-1}). Upon increasing reaction times, these bands diminished and a new band at 1005 - 1106 cm^{-1} appeared, attributed to C–O stretching of ozonides [17, 23]. Kogawa et al. and, other authors report that aldehyde bands were absent from the IR spectra of the ozonized oils [23, 26].

The FT-IR data for sunflower oil and ozonized sunflower oil of 24 hours shows absent bands of C=C–H stretching (3009 cm^{-1}) and appeared of C–O stretching of ozonides at 1106 cm^{-1}. ^{13}C NMR and Iodine Value described above confirmed the disappearance of double bonds [23].

The FT-IR spectrum of ethyl oleate before ozonolysis reaction presents two bands in this region at about 1117 and 1097 cm^{-1} attributable to the C-O stretching of the ester group. The strong ozonide band at about 1110 cm^{-1} gradually overlaps these two bands [18].

Oils are mainly formed of triglycerides, with different substitution patterns due to the length, degree and kind of unsaturation of the acyl groups, and by minor components such as mono- and di-glycerides. The chemical shifts of the triglycerides at ^{1}H NMR spectra are well known and presents the same standard for different vegetable oils [29].

Table 1. Ozononation reactions conditions

Author	Vegetable Oil	Sample/solvent	Gas Flow	Ozone Concentration	Time of reaction
Soriano et al., [20]	Sunflower Oil	50 mL oil 50 mL oil/400 mL deionized water	0,5 L/min	33-39 g/m^3	Until Solidification
Díaz et al., [21]	Sunflower Oil	150 g oil	42 L/h	79,5 mg/L	2 h
Díaz et al., [22]	Sunflower Oil	192g oil/20 mL of water	72 L/h	81.6 mg/L	3,5, 7 h
Díaz et al., [7]	Sunflower and Olive Oil	80 mL oil/8 mL water	30 L/h	75,2 mg/L	8,05 h (Sunflower Oil) 5,73 h (Olive Oil)
Moureu et al., [16]	Sunflower Oil	50 g oil 50g oil/5g ultra pure water	30 L/h	65 mg/L	1-7 h
Guerra-Blanco et al., [17]	Sunflower and Grape Oil	9 g each oil	0,5 L/min	30 mg/L	5 h (Until complete reaction)
Kogawa et al., [23]	Sunflower Flaxseed and Baru Oil	200 mL each oil Sunflower Oil/Water 9% (v/v)	1L/min	60 μg/mL	6, 12, 24 and 36 h 24h
Díaz and Gavín, [24]	Methyl Linoleato	1,6 mL/ 0,16 mL water	42 L/h	69 mg/L	7,25 min
Cataldo, [18]	Ethyl Oleato	130 mL	3,5 L/min	0.0260 mol/h	1, 2, 6, 8,10 and 15 h
Díaz et al., [25]	Coconut Oil	150 g 150g/15 mL of water or ethanol	54 L/h	37,5 mg/L	74,4 min
Zanardi et al., [26]	Sesame Oil	40 mL	1,5 L/min	55 mg/L	15, 30, 60, 90 e 120 min
Sega et al., [27]	Sesame Oil	40 mL	1,5 L/min	45 mg/L	15, 30, 45, 60, 75, 90, 105, and 120 min
Omonov et al., [28]	Canola Oil	100g/600 mL ethanol, methanol and ethyl acetate.	6,5 L/min	50 g/m^3	90 min

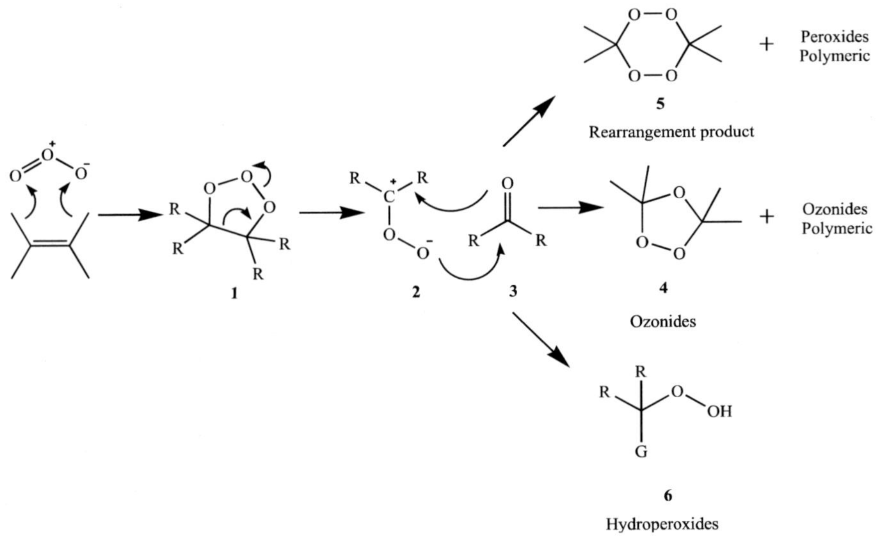

Scheme 1. Criegee mechanism for the reaction of ozone with carbon-carbon double bonds.

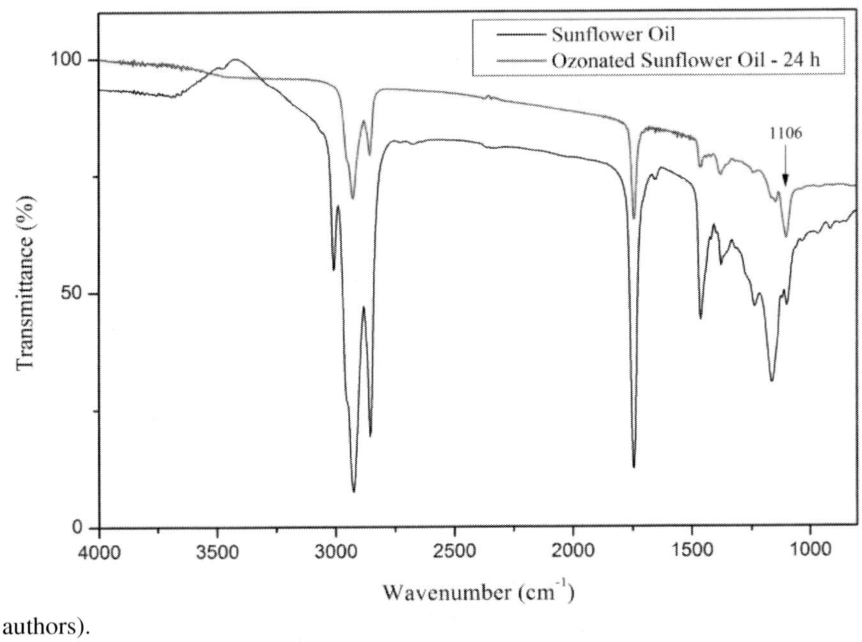

(Source: authors).

Figure 1. IR spectra of sunflower oil and ozonized sunflower oil – 24 hours.

In the ^1H NMR spectrum obtained from sunflower oil, the signs of olefinic hydrogens were observed in the region between 5.20 to 5.39 ppm. The signals at 2.00 and 2.74 ppm correspond to the protons of the methylene (-**CH$_2$**-CH=CH-), and the methylene group across the double bonds (-CH=CH-**CH$_2$**-CH=CH-), respectively. The double doublets at 4.10 and 4.26 ppm are assigned to hydrogens of methylene group *sn*-1 and 3 positions of the glycerol moiety (Figure 2), these signals remain in the ozonized oil spectra, indicating that there not changes occur in the glycerol moiety during the ozonolysis reaction [30]. The methyl group terminal presented a difference in the chemical shift depends of acyl groups, for linolenic acyl at 0.9 ppm and other acyl groups at 0.8 ppm [29].

The ^{13}C NMR spectrum shows two signals, at 172 and 173 ppm, refers to carbonyl carbons esters, sp^2 carbons corresponding to the unsaturation of the fatty acids were identified as the signals between 127 and 130 ppm. The signals 62 and 68 ppm refer to the carbon linked to the oxygen of the glycerol moiety, CH_2 and CH respectively, and the signals between 34 and 22 ppm refer to other carbon (-CH_2-) present in the structure [30]. The Table 2 shows mainly chemical shifts of ^1H and ^{13}C NMR for vegetable oils.

More specific information is obtained by ^1H and ^{13}C NMR spectroscopy in order to identify the changes in the chemical structure that occur during the reaction. Guerra Blanco et al., [17] observed in the ozonized oil spectral data, four different types of signals. These types are identified as conserved, decreasing, increasing, and increasing/decreasing signals [17].

The conserved signals are those that do not suffer any modification after treatment with ozone. These signals correspond to the α- and β-hydrogens of carbonyl groups, some methylene, and methyl of acyl groups and glycerol moiety. Therefore, protons that are not in proximity to double bonds not change its chemical shifts. Signals corresponding to the hydrogen of the double bonds (sp^2) (**CH=CH**), and to vinyl hydrogen (**CH_2**–C=C–**CH_2** and **CH_2**–C=C–CH_2–C=C–**CH_2**), (C=C–**CH_2**–C=C), show a decreasing after reaction [17].

New signal in ^1H NMR spectra of ozonized sesame oils at 5.15 ppm are attributed to the hydrogen of ozonide or 1,2,4 -trioxolane. Such evidence was confirmed by the appearance of a signal in ^{13}C NMR spectra at 104.33 ppm [26].

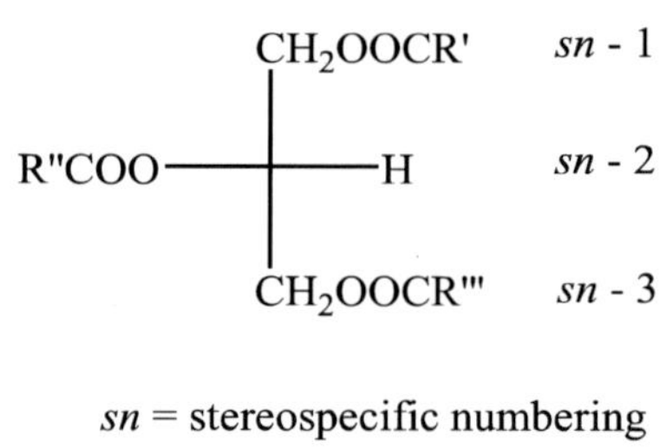

Figure 2. Glycerol moiety.

Table 2. ^1H and ^{13}C NMR data of vegetable oils (untreated) (Almeida et al. [30]; Sega et al. [27]; Guillén and Ruiz [29])

Moiety	^1H δ(ppm)	^{13}C δ(ppm)
–CH_3	0.90 – 0.80 (m)	14
– $(CH_2)_n$ –	1.40 – 1.15 (m)	22 – 32
–OCO–CH_2–CH_2–	1.70–1.50 (bs)	24
–OCO–CH_2–	2.35–2.20 (t)	33-34
–CH_2–CH=CH–	2.10–1.90 (m)	27
–CH=CH–CH_2–CH=CH–	2.80–2.70 (t)	25
–CH_2OCOR (*sn*-1 and 3 glycerol)	4.32 – 4.10 (dd, dd)	62
> CHOCOR (*sn*-2 glycerol)	5.26–5.20 (m)	68
–CH=CH–	5.40–5.26 (m)	127-130
–OCOR	--	172-173

bs: broad signal; m: multiplet; t: triplete, dd; doublet of doublet.

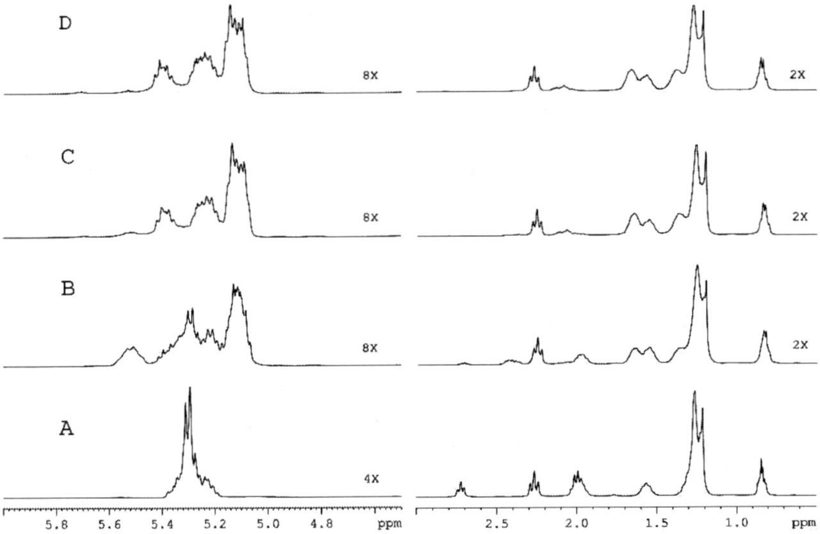

(Source: authors).

Figure 3. ^{1}H-NMR spectra expansions from 0 to 3.0 ppm and 4.5 to 6.0 ppm for sunflower oil ozonized for different exposure times: (A) 0 h; (B) 12 h; (C) 24 h; (D) 36 h. 2x, 4x and 8x: larger by factor.

Díaz et al., [21] observed in spectra of sunflower oil additional signals at 9.74 ppm and 9.63 ppm (triplets from aldehydic hydrogens), 5.6 ppm (olefinic protons signal could arise from hydroperoxides), 5.15 ppm (multiplet from ozonides). In ^{13}C NMR, aldehyde carbons resonating from 199 to 203 ppm, methynic carbons corresponding to ozonides and hydroperoxides from 104 to 122 ppm, methylenic carbons belong to ozonides and hydroperoxides from 42 to 44 ppm, and 23 to 24 ppm belong to methylenic carbons vicinal of ozonides [21].

Figure 3 shows the ^{1}H NMR spectra of ozonized sunflower oil at different times of reaction. Multiplet at 2.0 (–**CH$_2$**–CH=CH–), a triplet at 2.7 ppm (–CH=CH–**CH$_2$**–CH=CH–) and multiplet at 5.2 (–**CH=CH**–) present prior to reaction initiation, were absent from the spectra obtained after 24 and 36 h of ozonolysis. With the formation of ozonides, new signals appeared at 1.66 and 2.08 ppm, corresponding to methylene groups at the α-position and between ozonides, respectively. The multiplet at 5.5 ppm in the spectrum of ozonized sunflower oil 12h (B, Figure 2) and its absence from other spectra (C and D, Figure 2) indicate the formation of a homoallylic ozonide, with chemical shift of the signal for the remaining double bond hydrogens from 5.3 to 5.5 ppm. Table 3 shows the main ^{1}H and ^{13}C NMR chemical shifts for the vegetable oils [23, 30].

Physicochemical and Analytical Techniques

Some authors have remeasured the viscosity of sunflower [17], grape [17], sesame [26, 27] and coconut oil [25] and correlated viscosity increases with the ozonation degree increase. This observation is commonly explained by a decrease in the flexibility of ester chains because the total number of unsaturations has decreased, and by the formation of chemical species with higher molecular mass, which are assumed to be oligomers [25]. Modification of the unsaturated acyl chains ozonation kinetic affects the mobility and the reactivity of the species involved in the reaction. This aspect is of particular importance in terms of product characterization and evaluation of the ozonation reaction [26].

The ozonized coconut oil in ethanol showed lower viscosity. This behavior might be due to the solubility of oil with the ethanol. When water is used in the systems the viscosity increases because the emulsion appears due to the poor solubility of oil in water [25].

Table 3. ^1H and ^{13}C NMR data of ozonized oils. (Guerra Blanco et al., [17]; Sega et al., [27]; Kogawa et al., [23])

Moiety	^1H δ(ppm)	^{13}C δ(ppm)
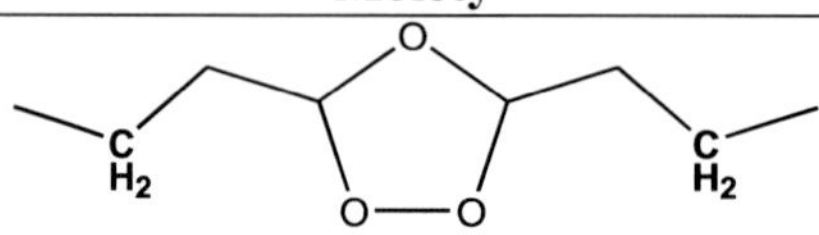	1.3 – 1.4 (bs)	23-24
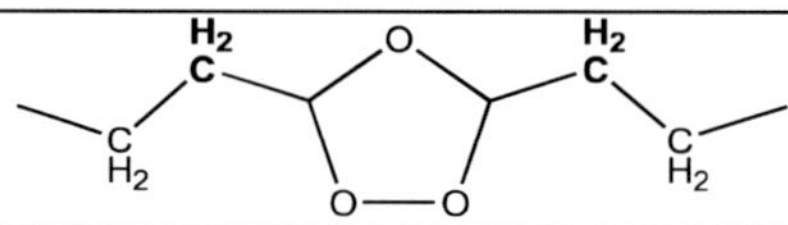	1.6 – 1.7 (m)	30
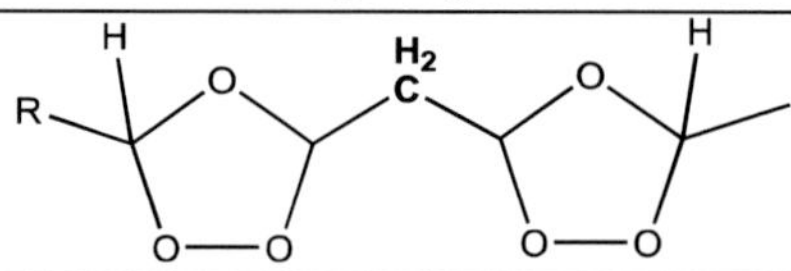	2,0 – 2.1 (m)	35 – 36
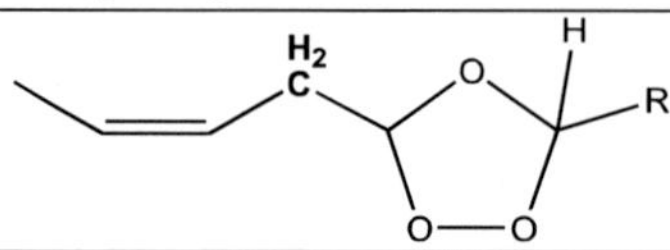	2.3 – 2.5 (m)	41 – 42
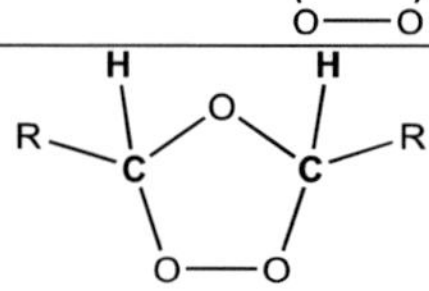 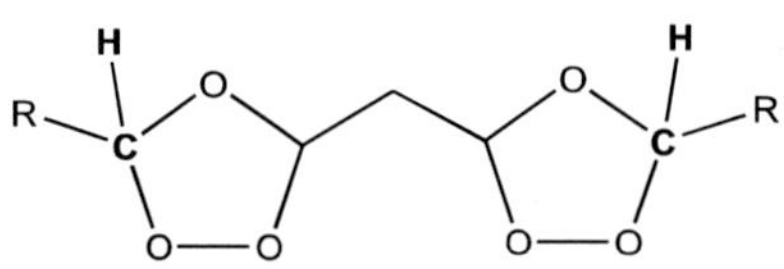	5.0 – 5.2 (m)	103 – 104
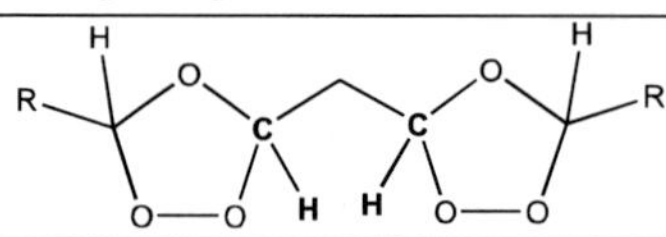	5.2 – 5,3 (m) 5.3 – 5.4 (m)	100
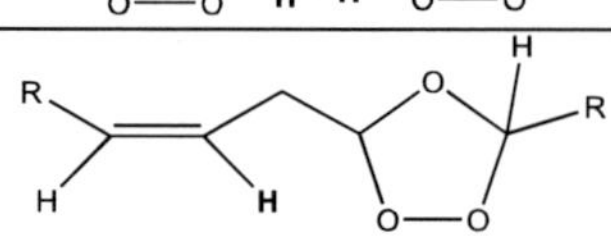 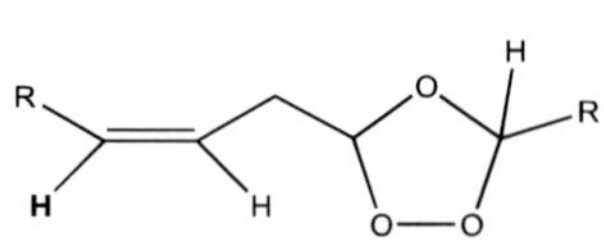	5.3 – 5,4 (m) 5,5 (m)	120 – 121 133 – 134

bs: broad signal; m: multiplet.

The study of the physicochemical properties of ozonized vegetable oils has a great importance for its characterization. Analytical methods such as peroxide, acidity, and iodine values usually carried out to follow-up the ozonization process and for determining the quality of ozonized vegetable oils [22].

The peroxide value (PV) represents the quantity of peroxide in the sample (meq kg^{-1}); acid value (AV) represents the present free acids (mg of KOH g^{-1}); and iodine value (IV) is a measure of total number of double bonds in the sample (g of iodide $(100\ g)^{-1}$ of sample). All these values are well described according to the European pharmacopoeia [31, 32] and Official Methods of Analysis of the Association of Official Analytical Chemists [33].

The literature describes an increase of acid and peroxide value and a decrease of iodine according increase of time of ozonation [16, 23, 7, 26, 27]. Acid value increases with the reaction time, indicating that acids can be formed through decomposition of ozonides or directly during the reaction. From the investigated samples, oleogel exhibited the highest acidity value, which can be explained by the formation of other peroxidic compounds in the presence of water, such as hydroxy-hydroperoxides [23]. According to Moureu et al., the increase seems to be slower for the samples ozonized without water, which is significantly higher for the oils ozonized with water [16].

It is well-known that iodine value is a measure of the total number of double bonds present in an oil sample [32]. The results also showed a decrease in iodine values with longer reaction times. The ozonized baru oil presented iodine value of 0.71 g $(100\ g)^{-1}$) after 24 hours of reaction. Nonetheless, flaxseed oil showed an iodine value of 44.61 g $(100\ g)^{-1}$ after 24 hours, indicating the presence of double bonds, as observed by the signals at 136–120 ppm in the ^{13}C NMR spectra [23]. Zanardi et al., reported a decrease of iodine value of 113.65 to 13.39 with the greater ozone dose used (9,900 mg) in Sesame oil [26].

The determination of the amount of peroxide present in oils ozonized is of fundamental importance because ozonides are responsible for the biological activity of these substances [7]. The peroxide value is expressed in milliequivalents (mEq) of active oxygen the quantity of peroxide present in 1000 g of sample, and commonly determined by iodometrics techniques, due to the ability of these compounds to oxidize iodide to iodine. Then, iodine is titrated with sodium thiosulfate solution [30].

According to the literature, for samples with high concentration of peroxides as ozonized oils, the experimental procedure described by American Oil Chemists' Society (A.O.C.S.) [33] becomes inadequate. The one-minute time, after addition iodide is insufficient for the peroxides present in the sample oxidize iodide to iodine. The dialkylperoxides and peroxides high molecular weight react slowly with iodide. Therefore, changes such as temperature and reaction time may be performed to obtain better results [34, 35].

Furthermore, the iodometrics methods used have some limitations. The two main sources of error in iodometrics tests are: (i) iodine reaction with the unsaturation of fatty acids and (ii) release of iodine from potassium iodide by oxygen in the solution to be titrated [36]. The determination of the amount of peroxide present in the ozonized oils is of great significance and new methodologies to quantify the peroxide must be developed.

Thermal Behavior (TG/DTG - DSC)

The TG curves (Figure 4) show mass losses for sunflower oil and ozonized oils. Sunflower oil (A) presented mass loss in one step in temperature between 300 and 450ºC and ozonated sunflower oils shows mass losses at three and four main steps, respectively. For oleogel (C), the first mass loss started at 70°C and for ozonized oil by 24 h (B) mass loss started at about 93ºC, revealing higher thermal stability of the latter.

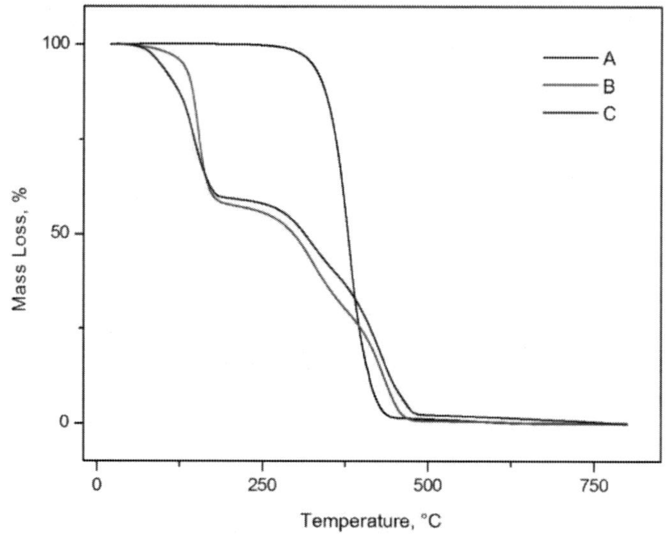

(1)

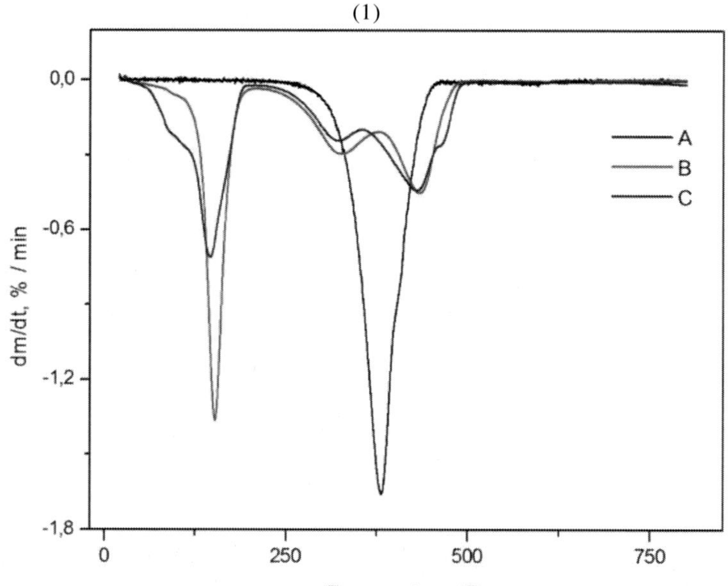

(2)

(Source: authors)

Figure 4. (1) TG (2) DTG curves for sunflower oils in a N_2 atmosphere (A) 0 h; (B) 24 h; (C) oleogel 24h.

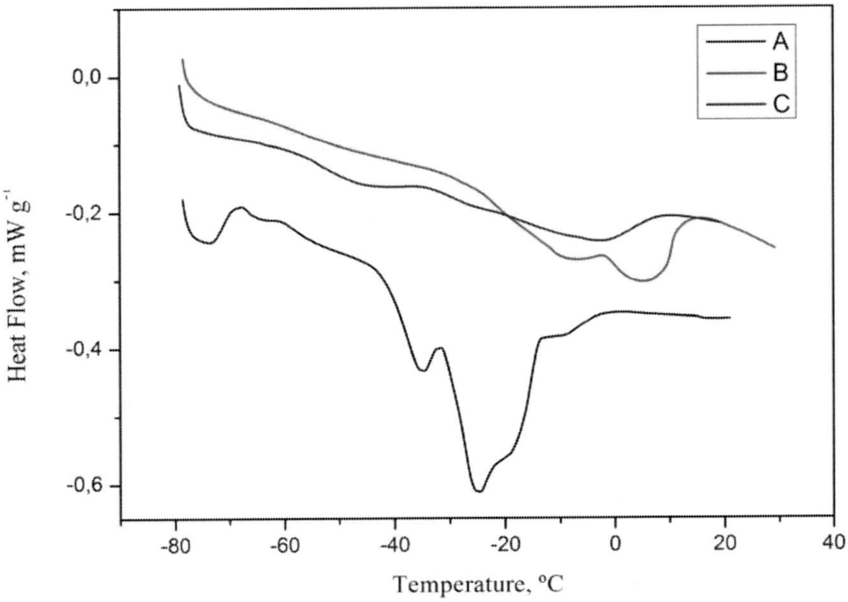

(Source: authors).

Figure 5. Heat DSC curves for sunflower oil: (A) 0 h; (B) 24 h; (C) oleogel 24h.

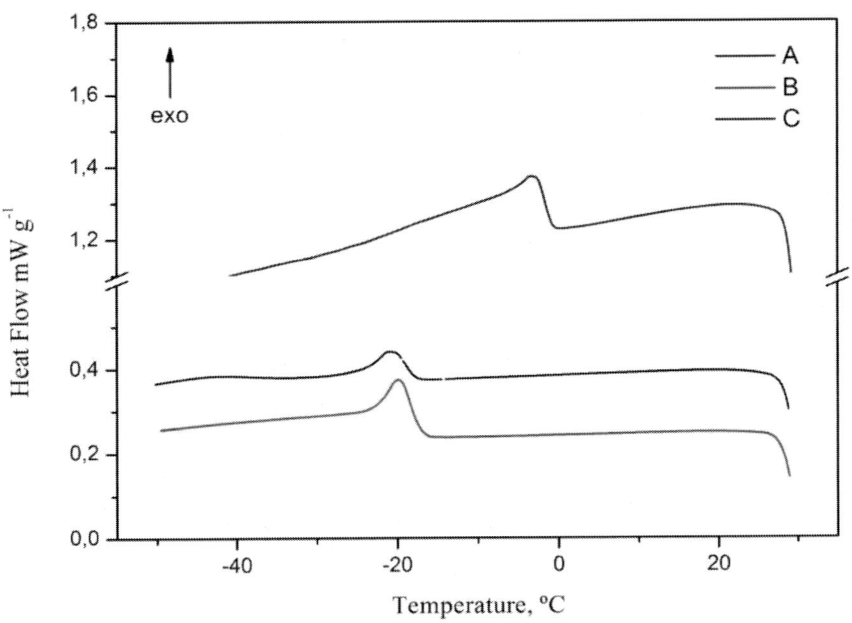

Figure 6. Cooling DSC curves for sunflower oil: (A) 0 h; (B) 12 h; (C) 24h.

Several samples of oils and ozonized oils were subjected to Differential Scanning Calorimetry (DSC) analysis. The study shows DSC curves corresponding by heating, cooling and degradation. DSC heat curve for sunflower oils (Figure 5) shows endothermic transitions corresponding to triglyceride (TAG) melting. Samples of sunflower oils show more than one endothermic transition to melting, suggesting polycrystalline state and the ozonized oil samples underwent endothermic transitions at higher temperatures than oils without ozonolysis. For the oleogel ozonized sunflower oil in presence of water was observed a Glass transition at about -56°C, which is a feature of polymeric materials or substances that undergo non-crystalline solidification. Glass transition was followed by an endothermic peak, corresponding to melting temperatures from -30 to 12°C and a fusion heat (ΔH) of 19.78 J g^{-1} [23].

Exothermic transitions observed in the cooling curve for DSC (Figure 6), corresponding to crystallization of fatty acids, sunflower oil exhibited transition characteristic of unsaturated acids, at -20.80°C. The ozonolysis leads to a decrease of unsaturated fatty acid decrease due to ozone reaction with the double bonds, leading the ozonized oils to present a similar thermal behavior to that of saturated fatty acids [23]. Melting and crystallization profiles from oil samples with high degrees of saturation involved higher temperature than for oils with high degrees of unsaturation [37].

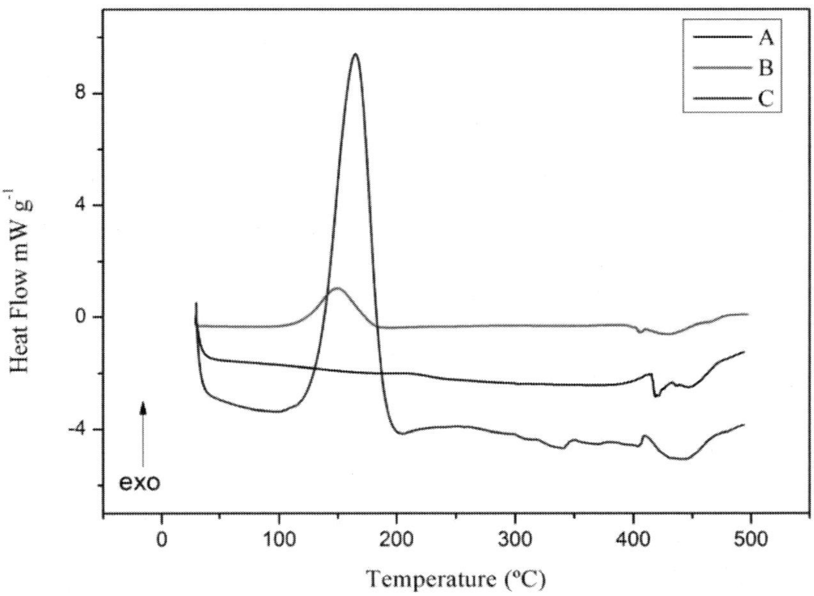

Figure 7. Thermal degradation DSC curves for sunflower oils. (A) 0 h; (B) 12 h; (C) 24h.

Cataldo and Kogawa et al. carried out a study of thermal behavior by Differential scanning calorimetry (DSC) [18, 23]. The ozonized vegetable oils, which seem completely stable at room temperature, should not be heated above 100 °C. The thermal analysis shows that the ozonides of ethyl oleate decompose showing a peak at about 155◦C and a broad exotherm. The accepted mechanism of the ozonide thermal decomposition involves the homolytic cleavage of O-O bond which is also considered to be the rate-determining step leading to aldehydes, carboxylic acid and other minor products [18]. Previously, Soriano et al. reported that ozonized sunflower oil decomposes at 152°C with a decomposition enthalpy of 878.7 J/g [38].

Kogawa et al. showed that thermal degradation by DSC and ozonized sunflower oil presented exothermic peaks attributed to degradation followed by oxidative decomposition of ozonides at about 150°C [23]. Oxidative decomposition was probably caused by oxygen released from ozonide degradation. The thermal analysis for oleogel exhibited exothermic peaks related to decomposition at similar ozonized sunflower oil of 24 h. Figure 7 shows thermal degradation DSC curves for sunflower oils.

Biological Activity

There is plenty of scientific literature dealing with the biological activities of ozonized vegetable oils, especially with respect to antimicrobial activity (antibiotic, antifungal and antiparasitic), as well as wound healing and antiulcer activities.

Pharmaceutical products based on ozonized vegetable oil are commercially available, mainly for dermatological treatments such as stimulants of wound healing and/or as disinfectants. Oleozon® (ozonized sunflower oil) is manufactured in Cuba; Cocozone in Britain by coconut oil ozonolysis; OOO and O2-ZAP are manufactured in Canada and the USA, respectively, from olive oil [39].

The commercial product Oleozon® was evaluated for its antimicrobial activity against resistant strains of methicillin-sensitive *Staphylococcus aureus* and *Staphylococcus epidermidis* [40], with MIC_{90} of 9.5 mg/mL at a time of action of 60 to 180 minutes. The ozonized sunflower oil was evaluated by Sechi et al. (2001) [41] as an antibiotic agent front various bacterial, sensitive and resistant strains of *Staphylococcus aureus*, *Enterococcus faecalis*, *Enterococcus faecium*, *Streptococcus pyogenes*, *Escherichia coli*, *Pseudomonas aeruginosa*, and various species of *Mycobacterium*. The results were very satisfactory, with MICs of 2.37 to 9.95 mg/mL for the species of mycobacteria, and 1.18 to 9.5 mg/ml for all other bacteria. At first, MIC values on this scale seem too big, but vegetable oils are complex mixtures of antioxidants and high-molecular-weight triacylglycerols, acting as a matrix capable of releasing active oxygen from ozonides, which have antimicrobial activity [41].

Previous works were followed by many researchers using other microorganisms. Additional parameters were analyzed starting from different vegetable oils. Table 1 summarizes this information.

It is interesting to note that many authors could relate the antimicrobial activity results to the peroxide index (PI) values showing that, to some extent, the higher the oil peroxide value, the greater the activity. After a certain point, an increase in PI does not influence the activity [16, 22, 25, 39, 43, 44]. This relationship suggests that the antimicrobial mechanism of action of ozonized vegetable oils is related to the action of oxidizing species, which was reported by Diaz et al. [49].

Studies show the influence of ozonation conditions and of the initial fatty acid composition of ozonized sunflower oils on their iodine index, peroxide index, acidity value and, antibacterial activity. The results indicated that fatty acid composition of the oils has no significant effect on the antimicrobial activity. However, the addition of water has a direct impact on the increase in peroxide index and so better antibacterial activity of oils ozonized with water [16, 23]. Recent results of our research group (not yet published) indicate that the neem oil ozonized in the presence of water also showed better antimicrobial activity. These results reveal opportunities for further studies and applications of ozonized oils.

Table 2. Ozonized vegetable oils and antimicrobial activity

Author	Product	Bacterial strain	MIC (mg/mL)/inhibition zone (mm)*/other information
Rodrigues et al. [42]	Bioperoxoil® (sunflower ozonized oil)	*S. aureus*, *E. coli*, *P. aeruginosa*, *Candida albicans* and *Salmonella typhimurium*	2.0-3.5 mg/mL
Díaz et al. [25]	Coconut ozonized oil (with or without protic solvent)	*S. aureus* and *P. aeruginosa*	17- 28.5 mm
Díaz et al. [22]	Sunflower ozonized oil (aqueous media)	*S. aureus*, *E. coli*, *P. aeruginosa*, *Bacillus subtilis*	9.5-38 mg/mL
Díaz et al. [7]	Sunflower and olive ozonized oils (aqueous media)	*S. aureus*, *E. coli*, *P. aeruginosa*, *B. subtilis*	0.95-14.25 mg/mL
Torres et al. [43]	Ozonized theobroma oil	*C. albicans*	5.78-93.75 mg/mL (depending on exposure time)
Gomez et al. [44]	Ozonized theobroma oil	*C. albicans*	2 – 15 mg/mL
Skalska et al. [39]	Sunflower ozonized oil	*E. coli*, *C. albicans*, *B. subtilis*	200-250 mg O_3/ g oil**
Daud et al. [45]	Bioperoxoil® (sunflower ozonized oil)	*Microsporum canis*	According to the authors, there is statistical indication that the ozonized oil acts against *M. canis* and there is clinical evidence of its action over this dermatophyte.
Menéndez et al. [46]	Oleozon® (sunflower ozonized oil)	phase III simple-blind study on onychomycosis patients	better therapeutic effect than topical ketoconazole
Gerrer et al. [47]	Bioperoxoil® (sunflower ozonized oil)	*C. albicans*, *Candida parapsiloses*, *Candida guilliermondii*, *Candida tropicalis*, *Tichosporum asaii*	11 – 19 mm
Ozyildis et al. [48]	Microcapsules containing ozonized red pepper seed oil	*E. coli*, *P. aeruginosa*, methicillin resistant *S. aureus*, vancomycin resistant *E. faecium*, *C albicans*	17 – 22.5 mm
Díaz et al. [49]	Sunflower ozonized oil (different sources for ozone formation)	*S. aureus*, *E. coli*, *P. aeruginosa*	4.75 – 28.5 mg/mL
Montececchi et al. [50]	Novox® (olive ozonized oil)	*S. aureus*, *Porphyromona gingivalis*	19.0 – 30.7 mm (0 – 1:128 dilutions)
Moureu et al. [16]	"Classical" and a "high oleic" sunflower oils ozonized with or without water	*S. aureus*, *E. coli*, *Streptococcus uberis*	5 - 40 mg/mL
Kogawa et al. [23]	Sunflower, flaxseed and baru ozonized oils (with or without water)	*S. aureus*, *E. coli*, *P. aeruginosa*, *E. faecalis*, oxacillin resistant *S. aureus*, vancomycin resistant *E. faecalis*	3 – 10 mg/mL

* MIC= minimal inhibitory concentration, using dilution methods; inhibition zone = using agar diffusion method.
** 100% of growth inhibition using the dose of 200 or 250 O_3 mg/g of oil during ozonolysis reaction.

There are also reports in the literature of antiparasitic activity of ozonized vegetable oils. Hernandez et al. evaluated the cytotoxic effect of Oleozon® on *Giardia duodenalis* trophozoites. The total cytotoxic effect on $15x10^4$ cells was obtained with 30μL (28.6 mg) of Oleozon® [51]. Direct cytotoxic-oxidant effect on the parasite *in vitro* may be one of the mechanisms of action for a parasitocidal effect of this product. Rajabi et al. studied the action of ozonized olive oil on *Leishmania major* promastigotes, the parasite that causes cutaneous leishmaniasis [52]. The authors found that IC_{50} were 120 mg/mL and 165 mg/mL for Glucantim and almost 2 and 3.5 μg/mL for ozonized olive oil, considering mean alive promastigotes determined by MTS method and light microscope [52].

Pai et al. described the potential of ozonized sesame oil to augment wound healing in rats. The animals were treated with two doses of ozonized sesame oil (peroxide values 500 and 700 mEq/1000 g, respectively), and framycetin, an antibacterial drug commonly used for treating wounds, was used as positive control [53]. Ozonized oil treated wounds had significantly higher tensile strength, collagen content and superoxide dismutase activity than that of the control treated wounds, which means that area treated with ozonized oil revealed better healing activity.

Some authors have reported other activities related to increasing on antioxidant enzyme activity. Rodriguez et al. [54] and Zamora et al. [55] reported the antiulcer activity of ozonized sunflower oil in rats. Rodriguez et al. evaluated the protective effect of ozonized oil on damage to digestive mucosa caused by ethanol action. Results demonstrate that OSO pretreatment exerts protective effects in ethanol-induced gastric ulcers in rats and these protective effects are mediated, at least partially, by stimulation of some important antioxidant enzymes such as SOD (superoxide dismutase) and GSH-Px (glutathione peroxidase), which are scavengers of ROS (reactive oxygen species) and therefore, prevent gastric injury induced by them [54]. Zamora et al. investigated the potential cytoprotective effects of ozonized sunflower oil in the damage of rat gastric mucosa induced by indomethacin. They also observed the cytoprotective effects of OSO in rat gastric mucosa. It was concluded that these cytoprotective effects are mediated somewhat by upregulation of the antioxidant system and mainly SOD [55].

Abu-Gharbieh et al. [56], similarly, evaluated the potential protective effect of ozonized olive oil in 2,4-dinitrobenzene sulfuric acid (DNBS) induced colitis in rats. Their results demonstrated that pretreatment with ozonized oil exerts protective effects in DNBS induced colitis in rats and provided evidence that protective the effects are mediated by stimulation of some antioxidant enzymes, once CAT (catalase), GSH-Px, and SOD activities were significantly increased in the distal colon of inflamed animals pretreated with olive ozonized oil with respect to control group dose dependently.

Sanchez et al. tested the antioxidant activity of a cosmetic formulation made with ozonized theobroma oil on rat skin irradiated with ultraviolet light. Again, the ozonized vegetable oil was able to stimulate the activity of antioxidant enzymes SOD and GSH-Px, which prevent skin injury induced by ultraviolet radiation [57].

Although ozonized oils have been used in many countries in the treatment of wounds and skin ulcers, reports on their cytotoxicity are scarce. Kogawa et al. [23] investigated sunflower and flaxseed ozonized oils (different ozonolysis times and conditions) for cytotoxicity against NIH/3T3 fibroblasts, but proved non-toxic when compared with doxorubicin, which is a desirable trait for the safe use of ozonized oils in patients. Also, authors evaluated the antitumor potential of the products against cancer cell lines: 786-0 (ATCC-CRL-1932, renal adenocarcinoma) HT-29 (ATCC-HTB-38, colon adenocarcinoma), MFC-7 (ATCC-HTB-22, breast adenocarcinoma), PC-3 (ATCC-CRL-1435, prostate adenocarcinoma), and B16-F10 (ATCC-CRL-6322, murine melanoma), and concluded that the oils were potentially active against neoplastic cell lines [23].

REFERENCES

[1] Swern, D. (1982). Bailey's Industrial Oil and Fat Products. New York, NY: John Wiley and Sons Inc.

[2] Fats and Oils Handbook (Nahrungsfette und Ole). (1998). Edited by Michael Bockisch. ISBN 978-0-9818936-0-0. 848p.

[3] Soriano, CR., Martinez, F. E. & Jorge, S. M. (2000). Cutaneous application of vegetable oil as a coadjutant in the nutritional management of preterm infants. *Journal of Pediatric Gastroenterology & Nutrition*, *31*, 387-390.

[4] Kumar, A., Aggarwal, G., Singh, K. & Harikumar, S. L. (2014). Comparison of vegetable and volatile oils as skin permeation enhancers for transdermal delivery of losartan potassium. *Der Pharmacia Lettre*, *6*, 199-213.

[5] Gullick, D. R., Pugh, W. J., Ingram, M. J., Cox, P. A. & Moss, G. P. (2009). Formulation and characterization of a captopril ethyl ester drug-in-adhesive-type patch for percutaneous absorption. *Drug Development and Industrial Pharmacy*, *36*, 926-32.

[6] Ledea, O. (2003). Ph.D Thesis, National Center for Scientific Reseach, Havana, Cuba.

[7] Díaz, M. F., Hernández, R., Martínez, G., Vidal, G., Gómez, M., Fernández, H. & Garcés, R. (2006b). Comparative Study of Ozonized Olive Oil and Ozonized Sunflower Oil. *Journal of Brazilian Chemical Society*, *17*, 403-407.

[8] Innovare. (2014). Ozonia 3000. Ozonized Sunflower seeds oil made in Innovares. Innovares® Knowledge in ozonized vegetable oils. Medical application of lipoperoxide and ozonide derived from ozonized oils. 26p. Avaliable from: http://www.innovares.com/download/ozonia/2015%20RazionaleOzonia3000_Libretto.pdf.

[9] Travagli, V., Zanardi, I., Valacchi, G. & Bocci, V. (2010). Ozone and ozonated oils in skin diseases: a review. *Mediators of Inflammation.*, Article ID 610418, 9 pages.

[10] Oleg, V. M., Claudia, N. K. & Irina, A. G. (2008) Ozone Therapy in Practice. Health Manual. Ministry of Health Service of the Russian Federation. The State Medical Academy of Nizhny, Novgorod, Russia, 42p. Avaliable from: https://www.absolute ozone.com/assets/ozone_ therapy_in_practice.pdf.

[11] Virginie, D., Sylvie, B., Michel, L., Jacques, F. & Michel, P. (2007). Fatty acid profiles of 80 vegetable oils with regard to their nutritional potential. *European Journal of Lipid Science and Technology*, *109*, 710–732.

[12] Zanardi, I., Burgassi, S., Paccagnini, E., Gentile, M., Bocci, V. & Travagli, V. (2013). What Is the Best Strategy for Enhancing the Effects of Topically Applied Ozonated Oils in Cutaneous Infections? *Biomed Research Inernational.* ID: 702949. 6p.

[13] Justyna, S., Bjarne, J., Espen, J., Rauno, F. & Lubomira Broniarz-Press. Jarl, B. R. (2008). Characterization of ozonated vegetable oils by spectroscopic and chromatographic methods. *Chemistry and Physics of Lipids.*, *151*, 85–91.

[14] Maritza, F. D., Frank, H., Oscar, L., José, A. & Gavín Sazatornil, Jesús Moleiro. (2003). ^{1}H NMR Study of Methyl Linoleate Ozonation. *Ozone: Science & Engineering.*, *25*, 121-126.

[15] Criegee, R. (1975). Mechanism of Ozonolisis, 14th ed., Angewandte Chemie Int: England.

[16] Moureu, S., Violleau, F., Haimoud-Lekhal, D. A. & Calmon, A. (2015). Ozonation of sunflower oils: Impact of experimental conditions on the composition and the antibacterial activity of ozonized oils. *Chemistry and Physics of Lipids*, *186*, 79-85.

[17] Guerra-Blanco, P., Poznyak, T., 18 Chairez, I. & Brito-Arias, M. (2015). Correlation of structural characterization and viscosity measurements with total unsaturation: An effective method for controlling ozonation in the preparation of ozonated grape seed and sunflower oils. *European Journal of Lipid Science and Technology*, *117*, 988-998.

[18] Cataldo, F. (2013). Chemical and thermochemical aspects of the ozonolysis of ethyloleate: Decomposition enthalpy of ethyl oleate ozonide. *Chemistry and Physics of Lipids.*, *175*, 41-49.

[19] Constantino, M. G. & Medeiros, E. F. (1988). Um gerador de fácil construção. *Química Nova*, *11*, 259-261.

[20] Soriano, N. U., Migo, V. P. & Matsumura, M. (2003). Ozonation of Sunflower Oil: Spectroscopic Monitoring of the Degree of Unsaturation. *Journal of the American Oil Chemists' Society.*, *80*, 997-1001.

[21] Dıaz, F. M., Sazatornil, J. A. G., Ledea, O., Hernandez, F., Alaiz, M. & Garces, R. (2005a). Spectroscopic Characterization of Ozonated Sunflower Oil. *Ozone: Science and Engineering*, *27*, 247–253.

[22] Díaz, M. F., Gavín, J. A., Gómez, M., Curtilles, V. & Hernández, F. (2006a). Study of ozonized sunflower oil using ^{1}H NMR and microbiological analysis. *Ozone: Science and Engineering.*, *28*, 59-63.

[23] Kogawa, N. R. A., Arruda, E. J., Micheletti, A. C., Matos, M. F. C., Oliveira, L. C. S., Lima, D. P., Carvalho, N. C. P., Oliveira, P. D., Cunha, M. C., Ojeda, M. & Beatriz, A. (2015). Synthesis, characterization, thermal behavior, and biological activity of ozonides from vegetable oils. *RCS Advances*, *5*, 65427-65436.

[24] Díaz, M. F. & Gavín, J. A. (2007). Characterization by NMR of Ozonized Methyl Linoleate. *Journal of Brazilian Chemistry Society*, *18*, 513-518.

[25] Díaz, M. F., Nunez, N., Quincose, D., Díaz, W. & Hernández, F. (2005b). Study of three systems of ozonized coconut oil. *Ozone: Science and Engineering*, *27*, 153-157.

[26] Zanardi, I., Travagli, V., Gabbrielli, A., Chiasserine, L. & Bocci, V. (2008). Physico-Chemical Characterization of Sesame Oil Derivatives. *Lipids*, *43*, 877-886.

[27] Sega, A., Zanardi, I., Chiasserinia, L. Gabbrielli, A., Bocci, V. & Travagli, V. (2010). Properties of sesame oil by detailed 1H and 13C NMR assignments before and after ozonation and their correlation with iodine value, peroxide value, and viscosity measurements. *Chemistry and Physics of Lipids*, *163*, 148–156.

[28] Omonov, T. S., Kharraz, E. & Curtis, J. M. (2011) Ozonolysis of Canola Oil: A Study of Product Yields and Ozonolysis Kinetics in Different Solvent Systems. *Journal of the American Oil Chemists' Society*, *88*, 689–705.

[29] Guillen, M. D. & Ruiz, A. (2001). High resolution ^{1}H nuclear magnetic resonance in the study of edible oils and fats. *Trends in Food Science & Technology*, *12*, 328–338.

[30] Almeida, N. R. (2013). Master's thesis. Síntese, caracterização e avaliação da atividade antimicrobiana de ozonídeos a partir de óleos vegetais. Universidade Federal do Mato Grosso do Sul, Campo Grande, MS, Brasil.

[31] European Pharmacopoeia. Acid value, Council of Europe, 5th edn, Strasbourg Cedex, France, 2004, pp. 127.

[32] European Pharmacopoeia. Iodine value, Council of Europe, 5th edn, Strasbourg Cedex, France, 2004, pp. 127–128.

[33] AOCS, Peroxide value. AOCS official method, in Official Methods and Recommended Practices of AOCS, ed. D. Firestone, American Oil Chemists' Society, Champaign, IL, 5th edn, 1998, p. 8.

[34] Richaud, E., Farcas, F., Fayolle, B., Audouin, L. & Verdu, J. (2006). Hydroperoxide titration by DSC in thermally oxidized polypropylene. *Polymer Testing.*, *25*, 829 – 838.

[35] Tellez, G. M., Lozano, O. L. & Gómez, M. F. D. (2006). Measurement of peroxide species in ozonized sunflower oil. *Ozone: Science and Engineering*, *28*, 181-185.

[36] Crowe, T. D. & White, P. J. (2001). Adaptation of the AOCS official method for measuring hydroperoxides from small-scale oil samples. *Journal of the American Oil Chemists' Society*, *78*, 1267-1269.

[37] Tan, C. P. & CheMan, Y. B. (2000). Differential scanning calorimetric analysis of edible oils: Comparison of thermal properties and chemical composition. *Journal of the American Oil Chemists' Society*, *77*, 143-155.

[38] Soriano, N. U., Migo, V. P. & Matsumura, M. (2005). Electrolytic cleavage of 1, 2, 4 trioxolanes of sunflower oil methyl esters. *Electrochimica Acta*, *50*, 1131–1137.

[39] Skalska, K., Ledakowicz, S., Perkowski, J. & Sencio, B., (2009). Germicidal properties of ozonized sunflower oil. *Ozone: Science and Engineering*, *31*, 232-237.

[40] Lezcano, I., Nunez, N., Espino, M. & Gomez, M., (2000). Antibacterial activity of ozonized sunflower oil, Oleozon, against *Staphylococcus aureus* and *Staphylococcus epidermidis*. *Ozone: Science and Engineering*, *22*, 207-214.

[41] Sechi, L. A., Lezcano, I., Nunez, N., Espim, M., Duprè, I., Pinna, A., Molicotti, P., Fadda, G. & Zanetti, S. (2001) Antibacterial activity of ozonized sunflower oil (Oleozon). *Journal of Applied Microbiology*, *90*, 279-284.

[42] Rodrigues, K. R., Cardoso, C. C., Caputo, L. R., Carvalho, J. C. T., Fiorini, J. E. & Schneedorf, M. (2004). Cicatrizing and antimicrobial properties of an ozonised oil from sunflower seeds. *Inflammopharmacology*, *12*, 261-267.

[43] Torres, I. F., Piñol, V. C., Urrutia, E. S. & Regueiferos, M. G. (2006). *In vitro* antimicrobial activity of ozonized theobroma oil againstcandida albicans. *Ozone: Science and Engineering*, *28*, 187-190.

[44] Gómez, M. F. D., Téllez, G. M., Sazatornil, J. A. G., Hernández, R. I. M. & Torres, I. F. (2007). Estudio de la grasa de teobroma ozonizada usando la Resonancia Magnética Nuclear protónica y análisis microbiológico. *Grasas Aceites*, *58*, 15-19.

[45] Daud, F. V., Ueda, S. M. Y., Navarini, A. & Mímica, L. M. J. (2011). The use of ozonized oil in the treatment of dermatophitosis caused by *Microsporum canis* in rabbits. *Brazilian Journal of Microbiology*, *42*, 274-281.

[46] Menéndez, S., Falcón, L. & Maqueira, Y., (2010). Therapeutic efficacy of topical OLEOZON® in patients suffering from onychomycosis. *Mycoses*, *54*, e272-e277.

[47] Gerrer, L. V., Cunha, K. C., Nogueira, M. C. L., Cardoso, C. C., Soares, M. M. C. N. & Almeida, M. T. G. (2012). *"In vitro"* antifungal activity of ozonized sunflower oil on yeasts from onychomycosis. *Brazilian Journal of Microbiology*, *43*, 1315-1318.

[48] Özyildiz. F., Karagönlü, S., Basal, G., Uzel, A. & Bayraktar, O. (2012). Microencapsulation of ozonized red pepper seed oil with antimicrobial activity and application to nonwoven fabric. *Letters in Applied Microbiology*, *56*, 168-179.

[49] Díaz, M. F., Sánchez, Y., Gómez, M., Hernández, F., Veloso, M. C., Pereira, P. A. P., Mangrich, A. S. & Andrade, J. B. (2012). Physicochemical characteristics of ozonized sunflower oils obtained by different procedures. *Grasas Aceites*, *63*, 466-474.
[50] Montevecchi, M., Dorigo, A., Cricca, M. & Cecchi, L. (2013). Comparison of the antibacterial activity of an ozonized oil with chlorhexidine digluconate and povidone-iodine. A disk diffusion test. *New Microbiologica*, *36*, 289-302.
[51] Hernández, F., Hernández, D., Zamora, Z., Díaz, M., Ancheta, O., Rodriguez, S. & Torres, D. (2009). *Giardia duodenalis*: Effects of an ozonized sunflower oil product (Oleozon®) on *in vitro* trophozoites. *Experimental Parasitology*, *121*, 208-212.
[52] Rajabi, O., Sazgarnia, A., Abbasi, F. & Layegh, P. (2015). The activity of ozonized olive oil against leishmania major promastigotes. *Iranian Journal of Basic Medical Sciences*, *18*, 915-919.
[53] Pai, S. A., Gagangras, S. A., Kulkarni, S. S. & Majumdar, A. S. (2014). Potential of ozonized sesame oil to augment wound healing in rats. *Indian Journal of Pharmaceutical Sciences*, *76*, 87-92.
[54] Rodríguez, Z. B. Z., Álvarez, R. G., Guanche, D., Merino, N., Rosales, F. H., Cepero, S. M., González, Y. A. & Schulz, S. (2007). Antioxidant mechanism is involved in the gastroprotective effects of ozonized sunflower oil in ethanol-induced ulcers in rats. *Mediators Inflammation*, 2007, 6 pages.
[55] Zamora, Z., Gonzáles, R., Guanche, D., Merino, N., Hernández, F., Alonso, Y. & Schulz, S. (2008). Ozonized sunflower oil reduces oxidative damage induced by indomethacin in rat gastric mucosa. *Inflammation Research*, *57*, 39-43.
[56] Abu-Gharbieh, E., Bayoumi, F. A. & Ahmed, N. G. (2014). Alleviation of antioxidant defense system by ozonized olive oil in dnbs-induced colitis in rat. *Mediators Inflammation*, 6 pages.
[57] Sánchez, Y., Díaz, M. F., Hernández, F., Gil, D. & García, G. (2012). Antioxidant effects of an ozonized theobroma oil formulation on damaged-inflammatory rat skin. *Grasas Aceites*, *62*, 105-110.

BIOGRAPHICAL SKETCH

Name: Nathália Rodrigues de Almeida
Affiliation: Institute of Chemistry of Federal University of Mato Grosso do Sul

Education:

- Ph.D student at Federal University of Mato Grosso do Sul (2013 to present)
- Master, degree in Chemistry, field of Organic Chemistry, Federal University of Mato Grosso do Sul (2013)
- Bachelor's degree in Pharmacy at Universidade Católica Dom Bosco (2008)

Address: Av. Senador Filinto Müller, 1555, CEP 79074-460 – Campo Grande, MS, Brazil

Research and Professional Experience: Organic Chemistry/Organic Synthesis

Professional Appointments:

- Junior Associate Editor of Orbital - The Electronic Journal of Chemistry (2014 to present)

Honors:

- **2012**: Finalist Project – SANTANDER Entrepreneurship - Prize of Biotechnology and Health.

Publications Last 3 Years:

1. Kogawa, N. R. A.; De Arruda, E. J.; Micheletti, A. C.; Matos, M. F. C.; De Oliveira, L. C. S.; De Lima, D. P.; Carvalho, N. C. P.; Oliveira, P. D.; Cunha, M. C.; Ojeda, M.; Beatriz, A. Synthesis, characterization, thermal behavior and biological activity of ozonides from vegetable oils. *RSC Advances: an international journal to further the chemical science*s, 2015, 5, 65427.

2. Almeida, N. R.; Beatriz, A.; Micheletti, A. C.; Arruda, E. J. Ozonized vegetable oils and therapeutic properties: A review. *Orbital: Electron. J. Chem*, 2013, 4, 313.

Name: Adilson Beatriz
Affiliation: Institute of Chemistry of Federal University of Mato Grosso do Sul (Brazil)

Education:

- Ph.D. degree in Science, field of Organic Chemistry from University of São Paulo – USP (2001).
- Bachelor's degree in Technological Chemistry by the Federal University of Mato Grosso do Sul (1995).

Address: Av. Senador Filinto Müller, 1555, CEP 79074-460 – Campo Grande, MS, Brazil

Research and Professional Experience: Organic Chemistry/Organic Synthesis

Experienced in Organic Synthesis, mainly in the following subjects: i) synthesis of resorcinolic lipids; ii) bioactive cage-like polycyclic compounds (development of new pharmaceuticals and new pesticides) and iii) Synthesis of polymers, surfactants etc. from biomass; iv) Chemical transformations of natural products; iv) Preparative biotransformations: whole cell and isolated enzymes in Organic Synthesis.

Professional Appointments:

- Post-Doctoral fellow, University of São Paulo (2001- 2002).
- Post-Doctoral fellow, University of Santiago de Compostela, USC, Spain (2009-2010).
- Research Associate Professor at Federal University of Mato Grosso do Sul (2006 to present).
- Research advisor in the Master and Doctor Programs in Chemistry (2006 to present).
- Head, Department of Chemistry, The Federal University of Mato Grosso do Sul (2009 to 2011).
- Editor-in-Chief of Orbital - The Electronic Journal of Chemistry (2009 to present).

Honors:

- **2015**: Finalist of the 7th Prize of Innovation Medical Services - Category Tropical Medicine, SANOFI
- **2014**: First Prize – 6th Prize of Innovative Medical Services – New Ways in Public Health, SANOFI and Portal Medical Services.

- **2012**: Finalist Project – SANTANDER Entrepreneurship - Prize of Biotechnology and Health.
- **2008**: First Prize - SANTANDER Regional Prize of Science and Innovation – Category: Biotechnology.

Publications Last 3 Years:

1. Schneider, B. U. C.; De Souza, A. M.; Beatriz, A.; Carvalho, P. C.; Mauro, M. O.; Karaziack, C. B.; De Lima, D. P.; Oliveira, R. J. Cardanol: toxicogenetic assessment and its effects when combined with cyclophosphamide. *Genetics and Molecular Biology*. In Press 2016.
2. Micheletti, A. C.; Honda, N. K.; Carvalho, N. C. P.; De Lima, D. P.; Beatriz, A. Design, Synthesis and *in vitro* Antimicrobial Activity Evaluation of Novel Hybrids of Lichexantone-THC Derivatives. *Orbital: Electron. J. Chem*. 2015, 7, 301.
3. Kogawa, N. R. A.; De Arruda, E. J.; Micheletti, A. C.; Matos, M. F. C.; De Oliveira, L. C. S.; De Lima, D. P.; Carvalho, N. C. P.; Oliveira, P. D.; Cunha, M. C.; Ojeda, M.; Beatriz, A. Synthesis, characterization, thermal behavior and biological activity of ozonides from vegetable oils. *RSC Advances: an international journal to further the chemical sciences*, 2015, 5, 65427.
4. Meza, A.; dos Santos, E. dos A.; Gomes, R. S.; De Lima, D. P.; Beatriz, A. Cytosporones: Isolation, Synthesis and Biological Activities of a Promising Class of Phenolic Lipids. *Current Organic Synthesis*, 2015, 12, 618.
5. Oliveira, R. J.; Navarro, S. D.; De Lima, D. P.; Meza, A.; Pesarini, J. R.; Gomes, R. S.; Karaziack, C. B.; Mauro, M. O.; Cunha-Laura, A. L.; Monreal, A. C. D.; Romão, W.; Lacerda Júnior, V.; Beatriz, A. A novel cytosporone 3-Heptyl-4,6-dihydroxy-

3H-isobenzofuran-1-one: synthesis; toxicological, apoptotic and immunomodulatory properties; and potentiation of mutagenic damage. *BMC Cancer*, 2015, 561, 1.
6. Araújo, Y. J. K.; Avvari, N. P.; Paiva, D. R.; De Lima, D. P.; Beatriz, A. Synthesis and enzymatic resolution of racemic 2,3-epoxy propyl esters obtained from glycerol. *Tetrahedron Letters*, 2015, 56, 1696.
7. Beatriz, A.; Gomes, R. S.; Constantino, M. G.; da Silva, G. V. J.; de OLIVEIRA, K. T. Synthesis of the Bicyclo[6.2.1] undecane Ring System by a Solvent-free Diels-Alder Reaction. *Tetrahedron Letters*, 2014, 55, 679.
8. Navarro, S. D.; Beatriz, A.; Meza, A.; Pesarini, J. R.; Gomes, R. S.; Karaziack, C. B.; Laura, A. L. C.; Monreal, A. C. D.; Romão, W.; Lacerda Junior, V.; Mauro, M. O.; Oliveira, R. J. A new synthetic resorcinolic lipid 3-Heptyl-3, 4, 6-trimethoxy-3H-isobenzofuran-1-one: evaluation of toxicology and ability to potentiate the mutagenic and apoptotic effects of cyclophosphamide. *European Journal of Medicinal Chemistry*, 2014, 132.
9. Rizzo, P. V. S.; Boarin, L. A.; Freitas, I. O. M.; Gomes, R. S.; Beatriz, A.; Rinaldi, A. W.; Domingues, N. L. C. The study of biocatalyzed thio-Michael reaction: a greener and multi-gram protocol. *Tetrahedron Letters*, 2014, 55, 430.
10. Polonini, H. C.; Lopes, R. S.; Beatriz, A.; Gomes, R. S.; Silva, A. O.; Lima, R. V.; Nunes, G. A.; Brandão, M. A. F.; Raposo, N. R. B.; De Lima, D. P. Synthesis and evaluation of octocrylene-inspired compounds for UV-filter activity. *Química Nova*, 2014, 37,1004.
11. Naujorks, A. A. S.; da Silva, A. O.; Lopes, R. S.; de Albuquerque, S.; Beatriz, A.; Marques, M. R.; de Lima, D. P. Novel naphthoquinone derivatives and evaluation of their trypanocidal and leishmanicidal activities. *Organic & Biomolecular Chemistry*, 2014, 13, 428.
12. Santos, E. A.; Prado, P. C.; Beatriz, A.; de Lima, D. P. Synthesis and biological activity of sulfur compounds showing structural analogy with combretastatin A-4. Química Nova, v. 36, p. 279-283, 2013.
13. da Silva, A. O.; LOPES, R. S.; de Lima, R, V.; Tozatti, C. S. S.; Marques, M. R.; de Albuquerque, S.; Beatriz, A.; de Lima, D. P. Synthesis and biological activity against Trypanosoma cruzi of substituted 1,4-naphthoquinones. *European Journal of Medicinal Chemistry*, v. 60, p. 51-56, 2013.
14. Micheletti, A. C.; Honda, N. K.; Pavan, F. R.; Leite, C. Q. F.; Matos, M. F. C.; Perdomo, R. T.; Bogo, D.; Alcântara, G. B.; Beatriz, A. Increment of Antimycobacterial Activity on Lichexanthone Derivatives. *Medicinal Chemistry (Hilversum)*, v. 9, p. 904-910, 2013.
15. Magalhães, H. I. F.; Wilke, D. V.; Bezerra, D. P.; Cavalcanti, B. C.; Rotta, R.; de Lima, D. P.; Beatriz, A.; de Moraes, M. O.; Diniz Filho, J.; Pessoa, C. O. (4-Methoxyphenyl) (3,4,5-trimethoxyphenyl) methanone inhibits tubulin polymerization, induces G2/M arrest, and triggers apoptosis in human leukemia HL-60 cells. *Toxicology and Applied Pharmacology*, v. 272, p. 117-126, 2013.

Name: Eduardo Jose de Arruda

Affiliation: Federal University of Grande Dourados

Education:

- Chemical Engineer – Industrial (FEI – 1984),
- MSc (1995) and PhD (1999) - Faculty of Chemical Engineering - FEQ/UNICAMP,

Address: Universidade Federal da Grande Dourados – UFGD / Faculdade de Ciências Exatas e Tecnologia - FACET – QUÍMICA - Rodovia Dourados - Itahum, km 12 - Cidade Universitária - C.P 533 - CEP: 79.804-970 - Dourados-MS

E-mails: ejarruda@gmail.com; eduardoarruda@ufgd.edu.br

Research and Professional Experience:

After PhD work on natural product, polymers, metal complexes, sustainable development, use of modified natural product or not, biomass and development of new products and technologies for medical applications and in health. Participated in university and classist boards from 2008 to 2014. It is a referee of national and international journals. The research group with researchers from UFGD, UFMS and UNICAMP has received recognition in the last seven years with the proposed use natural or modified products. Presently, his research focuses on the design and implementation of (bio) natural products, hybrid scaffolds and new strategies, metal-insecticides (metallo-insecticides) and photosensitizers for vector control of tropical diseases.

Professional Appointments:

- Full Professor at Universidade Católica Dom Bosco, Campo Grande, MS (1985 to 2008).
- Post-doctoral UNESP (2009/2010), FEQ / UNICAMP (2014) and UP - University of Porto - PT / INEB/i3S - Institute of Biomedical Engineering (2015).
- Adjunct Professor at School of Chemistry, Federal University of Grande Dourados (2008 to present).

Honors:

- **2014**: First Prize – 6th Prize of Innovative Medical Services – New Ways in Public Health, SANOFI and Portal Medical Services.

Publications Last 3 Years:

1. Silva, E. L.; Arruda, E. J.; Andrade, C. F. S.; Fernandes, M. F.; Teixeira, T. Z.; Scudeler, C. G. S.; Cabrini, I. Avaliação da Susceptibilidade de Populações de Aedes aegypti (Linnaeus) (Diptera: Culicidae) ao Inseticida Temephos nos Municípios de Maracaju e Naviraí, MS. *BioAssay (Piracicaba)*, v. 10, p. 1-5, 2015.

2. Catelan, T. B. S.; Arruda, E. J.; Oliveira, L. C. S.; Raminelli, C.; Cabrini, I.; Gaban, C. R. G.; Arakaki, A. H.; Nova, P. C. C. V. Evaluation of Toxicity of Phenolic Compounds Using Aedes aegypti (Diptera: Culicidae) and Artemia salina. *Advances in Infectious Diseases*, v. 05, p. 48-56, 2015.
3. Lescano, C. H.; Sanjinez-Argandoña, E. J.; Arruda, E. J.; Oliveira, I. P.; Baldivia, D. S.; Silva, L. R. Nutrients content, characterization and oil extraction from Acrocomia aculeata (Jacq.) Lodd. fruits. *African Journal of Food Science*, v. 9, p. 113-119, 2015.
4. Silva, A. K.; Arruda, E. J.; Fonseca, G. G.; Carvalho, C. T.; Silva, C. M.; Nova, P. C. C. V.; Gaban, C. R. G.; Cabrini, I. Evaluation of toxicity of Bordeaux Mixture in Aedes aegypti larvae (L. 1672) (Diptera: Culicidae) and Gram-negative and Gram-positive bacteria. *Journal of Mosquito Research*, p. 1-8, 2015.
5. Finoto, S.; Machulek Junior, A.; Caires, A. R. L.; Arruda, E. J.; Casagrande, G. A.; Raminelli, C.; Andrade, L. H. C.; Lima, S.M. New metalorgano-chalcogenide compounds based on polymeric frameworks constructed by Se-Hg intermolecular interactions: Preparation, structural characterization, and Raman evaluation. *Polyhedron*, v. 99, p. 96-102, 2015.
6. Kogawa, N. R. A.; De Arruda, E. J.; Micheletti, A. C.; Matos, M. F. C.; De Oliveira, L. C. S.; De Lima, D. P.; Carvalho, N. C. P.; Oliveira, P. D.; Cunha, M. C.; Ojeda, M.; Beatriz, A. Synthesis, characterization, thermal behavior and biological activity of ozonides from vegetable oils. *RSC Advances: an international journal to further the chemical sciences*, 2015, 5, 65427.
7. Carbonaro, E. S.; Arruda, E. J.; Oliveira, L. C. S.; Nova, P. C. C. V.; Arakaki, A. H.; Machulek Junior, A. A síntese de novos ativos larvicidas para o controle populacional do Aedes aegypti (Diptera: Culicidae). *Revista Brasileira de Inovação Tecnologica em Saúde*, v. 5, p. 10-26, 2015.
8. Scudeler, C. G. S.; Silva, T. G.; Fernandes, M. F.; Teixeira, T. Z.; Andrade, C. F. S.; Silva, I.; Arruda, E. J. Larval Susceptibility of Two Culex quinquefasciatus Populations (Diptera: Culicidae) Temephos® in the City of Naviraí, MS, Brazil. *Orbital: the Electronic Journal of Chemistry*, v. 7, p. 370-374, 2015.
9. Lima, A. R.; Arruda, E. J.; Cabrini, I.; Carvalho, C. T.; Fernandes, M. F.; Kato, M. F. H.; Andrade, C. F. S.; Silva, C. M. Insecticidal activity of Cu(II)-NTA in Aedes aegypti larvae (Diptera: Culicidae). *Orbital: the Electronic Journal of Chemistry*, v. 7, p. 369-375, 2015.
10. Gaban, C. R. G.; Dourado, D. M.; Silva, L. M. G. E.; Nova, P. C. C. V.; Cabrini, I.; Arruda, E. J. Morphological Changes in the Digestive System of Aedes aegypti L. Induced by [Cu(EDTA)]2- Complex Ions. *Journal of Mosquito Research*, v. 5, p. 1-9, 2015.
11. Andrade, R. C.; Almeida, C. F.; Suegama, P. H.; Arruda, E.J.; Arroyo, P. A.; Carvalho, C. T. Buriti palm stem as a potential renewable source for activated carbon production. *Environmental Technology & Innovation*, v. 3, p. 15-22, 2015.
12. Teixeira, T. Z.; Arruda, E. J.; Andrade, C. F. S.; Crispim, B. A.; Fernandes, M. F.; Silva, E. P.; Nakamura, A. K. S. Suscetibilidade de Larvas de Simulídeos ao Larvicida Temephos em Caarapó, MS. *BioAssay (Piracicaba)*, v. 9, p. 1-6, 2014.
13. Nardeli, J. V.; Arruda, E. J.; Carvalho, C. T.; Nova, P. C. C. V.; Cabrini, I.; Arakaki, A. H. Sintese, caracterização e atividade biológica do acetato de Cu(II) para larvas de

Aedes aegypti (Diptera: Culicidae) e bactérias Escherichia coli, Staphylococcus aureus, Salmonella typhimurium e Lysteria monocytogenes. *Orbital: the Electronic Journal of Chemistry*, v. 6, p. 122-129, 2014.
14. Lescano, C. H.; Sanjinez-Argandoña, E. J.; Arruda, E. J.; Kassuya, C. A. L.; Moraes, I. C. F. Acrocomia aculeata(Jacq.) Lodd. Oil Microencapsulation by Complex Coacervation: Preservation of Bioactive Compounds. *Journal of Encapsulation and Adsorption Sciences*, v. 04, p. 105-113, 2014.
15. Jacobowski, A. C.; Zobiole, N. N.; Padilha, P. M.; Moreno, S. E.; Arruda, E. J. Efeito Mutagênico do Edetato de Cobre ([Cu(EDTA)]-2) Livre e Nanoencapsulado em Camundongos e Peixes. *Journal of Brazilian Society of Ecotoxicology*, v. 8, p. 13-19, 2013.
16. Santos, G.; Arruda, E. J.; Oliveira, L. C. S.; Nova, P. C. C. V.; Ferreira, V. S. Desenvolvimento de metodologia eletroanalítica para determinação do antioxidante terc-butilhidroquinona (TBHQ) em amostras de biodiesel de soja. *Biofar: Revista de Biologia e Farmácia*, v. 9, p. 79-90, 2013.

Name: Ana Camila Micheletti

Affiliation: Institute of Chemistry of Federal University of Mato Grosso do Sul

Education:

- Ph.D. in Chemistry of Cerrado and Pantanal, field of Organic Chemistry from Federal University of Mato Grosso do Sul (2011).
- Master in Organic Chemistry (2007), from Federal University of Mato Grosso do Sul (2007).
- Bachelor's degree in Chemistry by the Federal University of Mato Grosso do Sul (2004).

Address: Av. Senador Filinto Müller, 1555, CEP 79074-460 – Campo Grande, MS, Brazil

Research and Professional Experience: working mainly in: chemistry of lichens and structural modification of natural products, organic synthesis, synthesis of bioactive molecules.

Professional Appointments:

- Professor at Federal University of Mato Grosso do Sul (2011 to present).
- Research advisor in the Master and Doctor Programs in Chemistry (2011 to present).

Publications Last 3 Years:

1. Micheletti, A. C.; Honda, N. K.; Carvalho, N. C. P.; De Lima, D. P.; Beatriz, A. Design, Synthesis and *in vitro* Antimicrobial Activity Evaluation of Novel Hybrids of Lichexantone-THC Derivatives. *Orbital: Electron. J. Chem.* 2015, 7, 301.
2. Kogawa, N. R. A.; De Arruda, E. J.; Micheletti, A. C.; Matos, M. F. C.; De Oliveira, L. C. S.; De Lima, D. P.; Carvalho, N. C. P.; Oliveira, P. D.; Cunha, M. C.; Ojeda, M.; Beatriz, A. Synthesis, characterization, thermal behavior and biological activity of ozonides from vegetable oils. *RSC Advances: an international journal to further the chemical sciences*, 2015, 5, 65427.
3. Micheletti, A. C.; Honda, N. K.; Pavan, F. R.; Leite, C. Q. F.; Matos, M. F. C.; Perdomo, R. T.; Bogo, D.; Alcântara, G. B.; Beatriz, A. Increment of Antimycobacterial Activity on Lichexanthone Derivatives. *Medicinal Chemistry (Hilversum)*, v. 9, p. 904-910, 2013.
4. Almeida, N. R.; Beatriz, A.; Micheletti, A. C.; Arruda, E. J. Ozonized vegetable oils and therapeutic properties: A review. *Orbital: Electron. J. Chem*, 2013, 4, 313.

Name: Dênis Pires de Lima

Affiliation: Institute of Chemistry of Federal University of Mato Grosso do Sul (Brazil)

Education:

- Bachelor's in Pharmacy from Universidade Federal de Minas Gerais (1986)
- Doctorate at Chemistry from Universidade Federal de Minas Gerais (1994).

Address: Av. Senador Filinto Müller, 1555, CEP 79074-460 – Campo Grande, MS, Brazil.

Research and Professional Experience: Organic Chemistry/Organic Synthesis/Medicinal Chemistry.

Experienced in the area of the molecular transformation of organic natural compounds, synthesis of phenolic compounds comprising naphthoquinones derivatives, analogues of resveratrol, combretastatin, and phenolic lipids as those isolated from cashew nut shell liquid (CNSL). Currently, biotransformation and bioremediation are exploited employing fungi isolated from many sources of the Brazilian biome.

Professional Appointments:

Part of Ph.D. program (1991-1992) was carried out at the UNIVERSITY OF ALBERTA (Edmonton-CANADA) working on the organic synthesis of antihypertensive agent, under supervision of the Professor Derrick L. J. Clive. The postdoctoral fellowship was taken at UNIVERSITY OF LIVERPOOL (Liverpool-UK) as a researcher of the group headed by Professor Stanley M. Roberts in the field of biotransformation.

Honors:

- **2015**: Finalist of the 7th Prize of Innovation Medical Services - Category Tropical Medicine, SANOFI
- **2014**: First Prize – 6th Prize of Innovative Medical Services – New Ways in Public Health, SANOFI and Portal Medical Services
- **2013**: Semifinalists Project – SANTANDER Prize of Science and Innovation
- **2005**: Best work in the 5th International Congress of Pharmaceutical Sciences, International Congress of Pharmaceutical sciences - Ribeirão Preto – SP - Brazil
- **2005**: Honored Mention, in the 5th International Congress of Pharmaceutical Sciences, International Congress of Pharmaceutical sciences - Ribeirão Preto – SP – Brazil

Publications Last 3 Years:

1. Oliveira, R. J.; Navarro, S. D.; De Lima, D. P.; Meza, A.; Pesarini, J. R.; Gomes, R. S.; Karaziack, C. B.; Mauro, M. O.; Cunha-Laura, A. L.; Monreal, A. C. D.; Romão, W.; Lacerda Júnior, V.; Beatriz, A. A novel cytosporone 3-Heptyl-4,6-dihydroxy-3H-isobenzofuran-1-one: synthesis; toxicological, apoptotic and immunomodulatory properties; and potentiation of mutagenic damage. *BMC Cancer*, 2015, 561, 1.
2. Meza, A.; dos Santos, E. dos A.; Gomes, R. S.; De Lima, D. P.; Beatriz, A. Cytosporones: Isolation, Synthesis and Biological Activities of a Promising Class of Phenolic Lipids. *Current Organic Synthesis*, 2015, 12, 618.
3. Micheletti, A. C.; Honda, N. K.; Carvalho, N. C. P.; De Lima, D. P.; Beatriz, A. Design, Synthesis and *in vitro* Antimicrobial Activity Evaluation of Novel Hybrids of Lichexantone-THC Derivatives. *Orbital: Electron. J. Chem*. 2015, 7, 301.
4. Carvalho, P.; Santos, E. A.; Catelan, B. U.; Matuo, R.; Pesarini, J.; Laura, A.; Monreal, A.; de Lima, D. P.; Oliveira, R. J.; Brochado, A. C. M.; Silva, A. Diaryl sulfide analogs of combretastatin A-4: toxicogenetic, immunomodulatory and apoptotic evaluations and prospects for use as a new chemotherapeutic drug. *Environmental Toxicology and Pharmacology*, 2015, 40, 715.
5. Naujorks, A. A. S.; da Silva, A. O.; Lopes, R. S.; de Albuquerque, S.; Beatriz, A.; Marques, M. R.; de Lima, D. P. Novel naphthoquinone derivatives and evaluation of their trypanocidal and leishmanicidal activities. *Organic & Biomolecular Chemistry*, 2014, 13, 428.
6. Araújo, Y. J. K.; Avvari, N. P.; Paiva, D. R.; De Lima, D. P.; Beatriz, A. Synthesis and enzymatic resolution of racemic 2,3-epoxy propyl esters obtained from glycerol. *Tetrahedron Letters*, 2015, 56, 1696.
7. Kogawa, N. R. A.; De Arruda, E. J.; Micheletti, A. C.; Matos, M. F. C.; De Oliveira, L. C. S.; De Lima, D. P.; Carvalho, N. C. P.; Oliveira, P. D.; Cunha, M. C.; Ojeda, M.; Beatriz, A. Synthesis, characterization, thermal behavior and biological activity of ozonides from vegetable oils. *RSC Advances: an international journal to further the chemical sciences*, 2015, 5, 65427.
8. Polonini, H. C.; Lopes, R. S.; Beatriz, A.; Gomes, R. S.; Silva, A. O.; Lima, R. V.; Nunes, G. A.; Brandão, M. A. F.; Raposo, N. R. B.; De Lima, D. P. Synthesis and

evaluation of octocrylene-inspired compounds for UV-filter activity. *Química Nova*, 2014, 37, 1004.
9. da Silva, A. O.; LOPES, R. S.; de Lima, R, V.; Tozatti, C. S. S.; Marques, M. R.; de Albuquerque, S.; Beatriz, A.; de Lima, D. P. Synthesis and biological activity against Trypanosoma cruzi of substituted 1,4-naphthoquinones. *European Journal of Medicinal Chemistry*, v. 60, p. 51-56, 2013.
10. Santos, E. A.; Prado, P. C.; Beatriz, A.; de Lima, D. P. Synthesis and biological activity of sulfur compounds showing structural analogy with combretastatin A-4. Química Nova, v. 36, p. 279-283, 2013.
11. Santos, E. A.; Hamel, E.; Bai, R.; Burnett, J. C. Tozatti, C. S. S.; Bogo, D.; Perdomo, R. T.; Antunes, A. M. M.; Marques, M. R.; MATOS, M. F. C.; de Lima, D. P. Synthesis and evaluation of diaryl sulfides and diaryl selenide compounds for antitubulin and cytotoxic activity. *Bioorganic & Medicinal Chemistry Letters* 2013, 23, 4669.
12. Magalhães, H. I. F.; Wilke, D. V.; Bezerra, D. P.; Cavalcanti, B. C.; Rotta, R.; de Lima, D. P.; Beatriz, A.; de Moraes, M. O.; Diniz Filho, J.; Pessoa, C. O. (4-Methoxyphenyl) (3,4,5-trimethoxyphenyl) methanone inhibits tubulin polymerization, induces G2/M arrest, and triggers apoptosis in human leukemia HL-60 cells. *Toxicology and Applied Pharmacology*, v. 272, p. 117-126, 2013.

Name: Lincoln Carlos Silva de Oliveira

Affiliation: Institute of Chemistry of Federal University of Mato Grosso do Sul

Education: Universidade Estadual Paulista (UNESP)

Address: Av. Senador Filinto Müller, 1555, CEP 79074-460 – Campo Grande, MS, BRAZIL

Research and Professional Experience: Analytical Chemistry / Thermal analysis

Professional Appointments:

- Ph.D. degree in Chemistry, field of Analytical Chemistry from Universidade Estadual Paulista – UNESP (1995).
- Research Associate Professor at Federal University of Mato Grosso do Sul (2007 to present).
- Research advisor in the Master and Doctor Programs in Chemistry (2007 to present).
- Principal, Institute of Chemistry, The Federal University of Mato Grosso do Sul (2013 to 2017).

Honors:

- **2015**: Finalist of the 7th Prize of Innovation Medical Services - Category Tropical Medicine, SANOFI.

- **2012**: Finalist Project – SANTANDER Entrepreneurship - Prize of Biotechnology and Health.

Publications Last 3 Years:

1. Cabral, M. R. P.; dos Santos, S. A. L.; Stropa, J. M.; da Silva, R. C. L.; Cardoso, C. A. L.; de Oliveira, L. C. S.; Scharf, D. R.; Simionatto, E. L.; Santiago, E.; Simionatto, E. Chemical composition and thermal properties of methyl and ethyl esters prepared from Aleurites moluccanus (L.) Willd (Euphorbiaceae) nut oil. *Industrial Crops and Products*, v. 85, p. 109-116, 2016.
2. Araujo, A. S. A.; Caramit, R. P.; de Oliveira, L. C. S.; Ferreira, V. S. Electroanalytical Method for Determining Pyrogallol in Biodiesel in the Presence of a Surfactant. *Electroanalysis*, v. 27, p. n/a-n/a, 2015.
3. Catelan, T. B. S.; Arruda, E. J.; Oliveira, L. C. S.; Raminelli, C.; Cabrini, I.; Gaban, C. R. G.; Arakaki, A. H.; Nova, P. C. C. V. Evaluation of Toxicity of Phenolic Compounds Using Aedes aegypti (Diptera: Culicidae) and Artemia salina. *Advances in Infectious Diseases*, v. 05, p. 48-56, 2015.
4. Pierezan, L.; Cabral, M. R. P.; Martins-Neto, D.; Stropa, J. M.; de Oliveira, L. C. S. et al. Chemical composition and crystallization temperatures of esters obtained from four vegetable oils extracted from seeds of Brazilian Cerrado plants. Química Nova, v. 38, p. 328-332, 2015.
5. Melnikov, P.; Arkhangelsky, I. V.; Nascimento, V. A.; Silva, A. F.; de Oliveira, L. C. S.; CONSOLO, L. Z.; Herrero, A. S. Thermolysis mechanism of dysprosium hexahydrate nitrate Dy(NO3)3·6H2O and modeling of intermediate decomposition products. *Journal of Thermal Analysis and Calorimetry*, v. xx, p. xx-xx, 2015.
6. Kogawa, N. R. A.; De Arruda, E. J.; Micheletti, A. C.; Matos, M. F. C.; De Oliveira, L. C. S.; De Lima, D. P.; Carvalho, N. C. P.; Oliveira, P. D.; Cunha, M. C.; Ojeda, M.; Beatriz, A. Synthesis, characterization, thermal behavior and biological activity of ozonides from vegetable oils. *RSC Advances: an international journal to further the chemical sciences*, 2015, 5, 65427.
7. Alexandre, E. C. F.; Silveira, E. V.; Castro, C. F. S.; Sales, J. F.; De Oliveira, L. C. S.; Viana, L. H.; Barbosa, L. C. A. Synthesis, characterization and study of the thermal behavior of methylic and ethylic biodiesel produced from tucumã (Astrocaryum huaimi Mart.) seed oil. *Fuel (Guildford)*, v. 161, p. 233-238, 2015.
8. Stropa, J.M.; Herrero, A. S.; Oliveira, S. C.; Cavalheiro, A. A.; Dantas, R. F.; Oliveira, S. L.; Machulek Junior, A.; de Oliveira, L. C. S.; Use of Natural Rubber Membranes as Support for Powder TiO 2 and Ag/TiO 2 Photocatalysts. *Journal of the Brazilian Chemical Society*, v. x, p. x, 2015.
9. Castro, D. C.; Cavalcante, R. P.; Jorge, J.; Martines, M. A. U.; de Oliveira, L. C. S.; Casagrande, G. A.; Machulek Junior, A. Synthesis and Characterization of Mesoporous Nb 2 O 5 and Its Application for Photocatalytic Degradation of the Herbicide Methylviologen. *Journal of the Brazilian Chemical Society*, v. 27, p. 303-313, 2015.
10. Armendáriz, V.; Martins, C. A.; Troiani, H. E.; de Oliveira, L. C. S.; Stropa, J. M.; Camara, G. A.; Martins, M. E.; Fernandez, P. S. Obtaining Clean and Well-dispersed Pt NPs with a Microwave-assisted Method. *Electrocatalysi*, v. 5, p. 279-287, 2014.

11. Lopes, S. A.; Cruz, N. A.; Manfroi, D. C.; Dias, R. G.; Silva, Margarete, S.; Zaghete, M. A.; dos Anjos, A.; Cavalheiro, A. A.; de Oliveira, L. C. S. Effect of the Iron Doping on the Thermal Decomposition of the Polymeric Precursor for the Titanium Dioxide Powder Synthesis. *Materials Science Forum*, v. 798-799, p. 211-216, 2014.
12. Melnikov, P.; Nascimento, V. A.; Arkhangelsky, I. V.; de Oliveira, L. C. S.; Silva, A. F.; Consolo, L. Z. Z. Thermogravimetric study of the scandium nitrate hexahydrate thermolysis and computer modeling of intermediate oxynitrates. *Journal of Thermal Analysis and Calorimetry*, v. 119, p. 1073-1079, 2014.
13. Ramos, D. D.; Bezerra, P. C. S.; Quina, F. H.; Dantas, R. F.; et al. Synthesis and characterization of TiO2 and TiO2/Ag for use in photodegradation of methylviologen, with kinetic study by laser flash photolysis. *Environmental Science and Pollution Research International*, v. 22, p. 774-783, 2014.
14. Melnikov, P.; Nascimento, V. A.; Arkhangelsky, I. V.; Consolo, L. Z. Z.; de Oliveira, L. C. S. Thermolysis mechanism of chromium nitrate nonahydrate and computerized modeling of intermediate products. *Journal of Thermal Analysis and Calorimetry*, v. 114, p. 1021-1027, 2013.
15. Melnikov, P.; Nascimento, V. A.; Arkhangelsky, I. V.; Consolo, L. Z. Z.; de Oliveira, L. C. S.. Thermal decomposition mechanism of iron(III) nitrate and characterization of intermediate products by the technique of computerized modeling. *Journal of Thermal Analysis and Calorimetry*, v. 115, p. 145-151, 2013.

In: Vegetable Oil
Editor: Brittany Holt

ISBN: 978-1-63485-128-2

Chapter 6

HEALTH BENEFITS OF VIRGIN COCONUT OIL

Gerard G. Dumancas[1,*], Lakshmi C. Kasi Viswanath[1], Arnie R. de Leon[2], Sindhura Ramasahayam[3], Randall Maples[4], Rangika Hikkaduwa Koralege[5], Undugodage Don Nuwan Perera[6], Joel Langford[1], Aamina Shakir[7] and Samuel Castles[1]

[1]Department of Chemistry, Oklahoma Baptist University, Shawnee, OK, US
[2]Department of Chemistry, Vanderbilt University, Nashville, TN, US
[3]Department of Hematology and Genetics, Children's Hospital of Philadelphia, Philadelphia, PA, US
[4]Science and Mathematics, Eastern Oklahoma State College, Wilburton, OK, US
[5]School of Chemical Engineering, Oklahoma State University, Stillwater, OK, US
[6]Department of Chemistry, Oklahoma State University, Stillwater, OK, US
[7]College of Medicine, University of Oklahoma Health Sciences Center, Oklahoma City, US

ABSTRACT

Virgin coconut oil (VCO) is a product that can be produced from fresh coconut meat, milk, or residue. Over the years, it has become known as a popular functional food oil. It is considered to be the newest, high-value coconut product, very much sought for its human, nutraceutical benefits, as well as a functional food. Its increasing popularity can be attributed to numerous studies showing its beneficial effects. Several studies have investigated the pharmacological properties of VCO including anti-inflammatory, analgesic, antipyretic, anti-oxidant, anti-stress, and antimicrobial properties. Furthermore, other studies have also investigated the bone loss prevention as well as cardioprotective effects of VCO. For example, administration of VCO in animal studies (i.e., Sprague-Dawley rats) showed significant antithrombotic effect compared to copra oil. The effects were comparable with sunflower oil fed animals. This chapter will discuss the chemical

[*]Corresponding author: Email: gerard.dumancas@okstate.edu, Phone: +14057308752.

properties, sources, synthesis, preparation, uses, and worldwide production of VCO. A particular focus will be on the health benefits of VCO including its most recent findings.

ABBREVIATIONS

AA	Arachidonic acid
APCC	Asian Pacific Coconut Community
APP	Acute-phase protein
BAFPS	Bureau of Agriculture and Fisheries Product Standards
BM	Bawalan-Masa
COX-2	Cyclooxygenase-2
CMV	Cytomegalovirus
DCN	Desiccated Coconut
DME	Direct Micro Expelling
EBV	Epstein-Barr Virus
FA	Fatty Acid
FFA	Free fatty acid
GAE	Gallic acid equivalents
GPX	Glutathione peroxidase
GSH	Glutathione
HDL	High-density lipoprotein
ICS	International Certification Services
IL-6	Interleukin-6
iNOS	Inducible nitric oxide synthase
LDL	Low-density lipoprotein
MCFA	Medium chain fatty acid
MCT	Medium chain triglyceride
MDA	Malondialdehyde
MIC	Minimum inhibitory concentration
NO	Nitric oxide
NSAID	Non-steroidal inflammatory drug
PG	Prostaglandin
PNS	Philippine National Standard
P. acnes	*Propionibacterium acnes*
RA	Rheumatoid arthritis
RBD	Refined, bleached, deodorized
SFA	Saturated fatty acid
SOD	Superoxide dismutase
TNF-α	tumor necrosis factors-α
VCO	Virgin coconut oil

1. INTRODUCTION

Coconut oil is one of the most important food oils in the world as a source of dietary fat. It is extensively used as cooking oil, and also in food industries such as confectionary and baking products [1]. There are two types of coconut oil based on how they are obtained from the coconut meat: refined, bleached and deodorized coconut oil (RBD) and virgin coconut oil (VCO). RBD oil is made from copra, which is the dried coconut kernel or meat, produced through smoke drying, sun drying or a combination of both. Then the clean, ground and steamed copra is pressed to obtain the coconut oil. Although coconut is one of the healthiest foods known to mankind, the drying process and unhygienic storage and handling makes the extracted oil unsafe for human consumption. Therefore the refining, bleaching, and deodorizing process is necessary after the extraction. This process requires heating the oil at high temperatures, between 204°C and 245°C, which destroys the essential amino acids, tocopherols (i.e., Vitamin E) and other valuable compounds present in coconut oil [2-6].

Coconut oil can also be extracted from fresh coconut kernels, the process of which does not involve high temperature treatments. The Philippine national standard defines VCO as the oil obtained from the fresh, mature kernel (meat) of the coconut by mechanical or natural means, with or without the use of heat, without undergoing chemical refining, bleaching or deodorizing processes, which does not lead to the alteration of the nature of the oil [5]. This is the purest form of coconut oil, which has the fresh coconut aroma and contains natural Vitamin E and other valuable compounds present in the coconut meat. VCO has a clear water appearance. RDB oil, in contrast, can appear yellow, pink or red-orange due to the contaminants (microbial or other), or high temperature processing. VCO has low free fatty acid content and peroxide value since it does not undergo the copra-making process, which eliminates atmospheric and hydrolytic oxidation compared to the RBD process [5].

There is a growing demand for VCO in United Sates and other developed countries, which can be attributed to the increasing number of books, journal articles and other scientific literature published on the health benefits of VCO. The amount of VCO manufactured in tropical countries has grown rapidly in the last two decades. For example, the number of VCO producers increased from 20 in 2003 to 200-300 producers in 2005 in the Philippines [7]. The exported VCO from the Philippines has increased from 19 metric tons to 177 metric tons from 2002 to 2004 as reported [8]. It was reported in another document published by the Philippine Coconut Authority that the export of VCO increased from 2737 metric tons to 6002 metric tons from 2010 to 2012 showing that the demand and the production of VCO is rapidly increasing [9]. Worldwide production of VCO shall further be discussed in the succeeding sections.

2. CHEMICAL PROPERTIES OF VCO

VCO, which is extracted directly from coconut milk by a wet process under controlled temperature conditions, retains more of its beneficial components than copra oil. Particularly, this extraction process avoids the loss of minor components such as pro-vitamin A, vitamin E and polyphenols due to UV irradiation from sunlight during the drying of copra [4, 3, 10]. VCO has a rich content of medium chain fatty acids (MCFAs), predominantly lauric acid;

others include caproic acid, caprylic acid and capric acid [10]. A study conducted by Mansor et al., [10] on VCO extracted by different processing methods reported that the lauric acid contents ranged from 46.36%-48.42% and the total MCFA in the oil (caproic acid, caprylic acid, capric acid and lauric acid) ranged from 59.02% to 62.27% of the total fatty acids. According to his findings, the highest lauric acid content was reported from the samples extracted using the fermentation process, followed by fresh-dry, chilling, and enzyme methods.

Apart from the high concentration of MCFAs, VCO also contains saturated fatty acids (SFAs) such as myristic acid, palmitic acid, and stearic acid, as well as unsaturated fatty acids, both mono-and di-unsaturated fatty acids. As reported by Mansor et al., [10] the concentrations of SFA and total unsaturated fatty acids ranged from 28% to 31% and 6.73% to 8.13%, respectively. Table 1 presents the fatty acid compositions of VCO produced from different methods and the Asian Pacific Coconut Community (APCC) standard compositions for VCO [10]. Variations in fatty acid composition could result from different methods of processing among VCO samples (Table 1) [3, 10, 11].

Iodine value, saponification value, and peroxide content are some of the most important chemical properties that are vital in characterizing the quality of VCO. Other important physicochemical properties in VCO include free fatty acid composition, moisture content, and viscosity (Table 2).

Table 1. Fatty acid (FA) composition of VCO produced from different methods and APCC standard compositions for VCO (% area)

FA	Extraction Method				APCC Standard
	Chilling	Enzyme	Fermentation	Fresh-dry	
C6 (caproic acid)	0.57 ± 0.00	0.52 ± 0.00	0.57 ± 0.01	0.55 ± 0.00	0.40-0.60
C8 (caprylic acid)	7.39 ± 0.03	6.63 ± 0.01	7.21±0.13	7.23 ± 0.00	5.00-10.00
C10 (capric acid)	6.15 ± 0.01	5.49±0.00	6.07±0.10	5.94 ± 0.01	4.50-8.00
C12 (lauric acid)	48.05 ± 0.11	46.36±0.00	48.42 ± 0.90	48.07 ± 0.02	43.00-53.00
C14 (myristic acid)	18.45 ± 0.03	19.54±0.01	18.75 ± 0.34	19.23 ± 0.00	16.00-21.00
C16 (palmitic)	8.94 ± 0.05	9.94 ± 0.01	9.06 ± 0.16	8.91 ± 0.01	7.50-10.00
C18 (stearic acid)	2.96 ± 0.03	3.37 ± 0.00	3.15 ± 0.00	3.17 ± 0.09	2.00-4.00
C18: 1 (oleic acid)	6.18 ± 0.03	6.50 ± 0.01	6.35 ± 0.01	5.79 ± 0.01	5.00-10.00
C18: 2 (linoleic acid)	1.31 ± 0.01	1.63 ± 0.00	1.36 ± 0.00	1.12 ± 0.00	1.00-2.50

Table 2. Physicochemical analysis of VCO extracted using the different methods [10]

Analysis	Extraction Method				APCC Standard, 2007
	Chilling	Fermentation	Fresh-dry	Enzyme	
Iodine value ($g\ I_2/100\ g$ fats)	4.13 ± 0.02	4.30 ± 0.07	4.18 ± 0.04	4.26 ± 0.05	4.10
Free fatty acid (mg KOH/g oil)	0.31 ± 0.01	0.29 ± 0.02	0.46 ± 0.01	0.35 ± 0.01	0.5 max
Saponification value (mg KOH/g oil)	258.23 ± 3.09	256.73 ± 0.85	258.42 ± 1.41	262.72 ± 0.32	250-260 min
Moisture content (% wt)	0.11 ± 0.01	0.06 ± 0.00	0.04 ± 0.00	0.11 ± 0.01	0.1-0.5
Viscosity (Pa.s)	48.93 ± 0.31	48.73 ± 0.46	50.93 ± 0.31	48.93 ± 0.31	NA

Iodine value of VCO gives an indication of its saturation level. The low content of iodine value as described by Mansor et al. (Table 2) [10] and Marina et al. [3] verified that VCO has high degree of saturation. Consequently, VCO has high resistance to oxidative rancidity. Free fatty acids (FFAs) are responsible for the undesirable flavor and aromas present in VCO and are formed by the hydrolytic rancidity, due to ester hydrolysis either by lipases or moisture. Measured FFA contents were varied from 0.29-0.46 (mg KOH/g oil) in a study conducted by Mansor et al. [10].

The saponification value, another important chemical characteristic, measuresthe average molecular weight of all the fatty acids present in VCO. One study found out that VCO has very high saponification values compared to other vegetable oils (3). The higher the saponification value, the shorter the fatty acids on the glycerol backbone, indicating that VCO contains a higher amount of short-chain fatty acids. Mansor et al. reported the saponification values from four different extraction methods (Table 2) [10].

VCO can also be characterized by the presence of antioxidants. Tocopherols, which are natural lipophilic antioxidants, are known to be found in vegetable oils including VCO. Mansor et al. [10] detected three types of tocopherols present in VCO including beta, gamma and delta forms. Beta-tocopherol ranged from 0.04 – 0.05 mg/kg, gamma-tocopherol ranged from 0.01–0.05 mg/kg, and delta-tocopherol was detected at a very low concentration levels (1.30 x 10-5 to 1.10 x 10-3 mg/kg) in VCO.

VCO is also good source of phenolic compounds, which are potential natural antioxidants found in foods. Total phenolic contents present in VCO will vary based on coconut varieties and oil extraction processes [4, 3, 10]. VCO has been shown to have a high total phenolic content (11.82–29.18 mg gallic acid equivalents [GAE]/100g oil), which is responsible for its high antioxidant properties (antioxidant activity ranging from 52.54% to 79.87%) [12]. Another study conducted by Arlee et al. has reported total phenolic contents of VCO ranged from 48.17–57.89 mg GAE/100 g oil [13].

Moisture content is another important quality characteristics for oils and fats. Low moisture levels will increase the shelf life by preventing oxidation and rancidity processes, whereas high moisture content will assist in hydrolysis. Mansor et al. reported the moisture content (% wt) and viscosities for VCO extracted using four extraction methods (Table 2) [10]. The highest recorded viscosity was from fresh-dry method while the lowest was from fermentation method.

Besides the abovementioned properties, formation of peroxides and hydroperoxide in the initial stage of lipid oxidation is also an important property of VCO that should be noted, because it reflects the tendency of the oil to become rancid. As observed by Marina et al. [3] peroxide values ranged from 0.21 to 0.63 mequiv oxygen/kg oil, which were far below the maximum limits according to Codex standard [14].

3. Sources, Synthesis, Preparation, and Uses of VCO

3.1. Sources

Coconut oil extracted from the fruit of the coconut palm (*Cocos nucifera L.*) is of two different types: (i) coconut or copra oil obtained from dry coconut flesh and (ii) VCO acquired from fresh coconut flesh. VCO can either be obtained directly from the fresh comminuted (grated, chopped, granulated) coconut meat or from coconut milk or from coconut milk residue [5]. However, the choice of the technology to be implemented for VCO processing depends on various parameters such as scale of operations, the degree of mechanization desired, the amount of investment available and the demands of the prospective buyer [5].

3.2. Processing of VCO

The Philippine National Standard (PNS) for VCO (PNS/The Bureau of Agriculture and Fisheries Product Standards (BAFPS) 22:2004/International Certification Services (ICS) 67.2000.10) defines it as: "oil obtained from the fresh, mature kernel (meat) of the coconut by mechanical or natural means, with or without the use of heat, without undergoing chemical refining, bleaching or deodorizing, and which does not lead to the alteration of the nature of the oil. VCO is essentially water-clear or colorless. It contains natural vitamin E and has not undergone any hydrolytic and atmospheric oxidation as demonstrated by its very low free-fatty acid content (even without refining) and low peroxide value" [15]. VCO has a fresh coconut aroma and its intensity depends on the extraction process. Hence, unlike refined coconut oil, the production process of VCO does not go through the RBD procedures, which are carried out at high temperature between 204°C and 245°C as mentioned earlier [16]. VCO can be produced even at home without the need for any specialized equipment. As a means to ensure the production of high quality oil, the VCO processor used at home or plant should strictly comply with the good manufacturing practices and quality control procedures [17].

The production of VCO is divided into three stages, (i) pre-processing, (ii) processing and (iii) post processing.

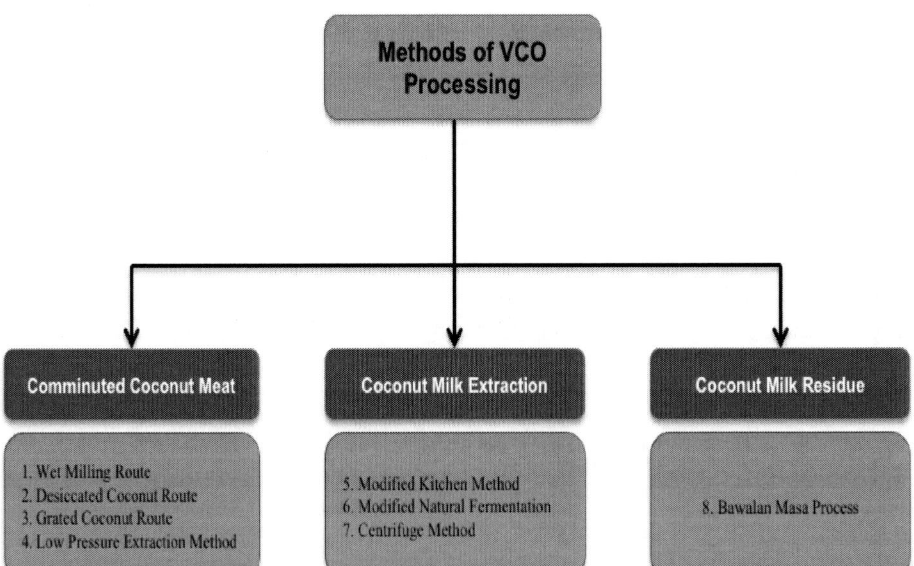

Figure 1. Methods of VCO processing.

3.2.1. Pre-processing Stage

The pre-processing stage involves all the steps performed before the opening of the fresh coconut such as on-farm activities (harvesting, collection and husking of nuts), transport from the farm to the VCO processing site (factory or home), storage, and selection for daily processing [18].

3.2.2. Processing Stage

The processing stage begins with the opening of the fresh kernel and is complete until the final recovery of VCO. The VCO production process can be carried out in eight different ways based on the desired final quality of oil [5, 4, 19-20]. The common method of classification is based on whether a dry or wet process is employed. However, methods can also be classified based on the precursor form of coconut used (Figure 1).

A. Fresh Coconut Meat

(i) High Pressure Expelling Process

The high pressure expelling process is considered to be the most appropriate method to process a capacity of 3000 or more coconuts a day [5]. The method demands the use of mechanical dryers and high-pressure expellers that are equipped with water-cooled shafts. VCO obtained from the high-pressure expeller process is highly viscous as it contains all of the natural gums present in the parent fresh kernel. There are three different types of high-pressure expeller method of VCO production based on the mode of preparation of kernels prior to drying, the Wet Milling Route, the Desiccated Coconut Route, and the Grated Coconut Route [5, 17].

- **Wet Milling Route**. In the wet milling route, the de-shelled meat is first washed with copious amount of water and then milled. The particulated coconut meat is then dried to 75°C and the oil is extracted using a screw-type press to produce VCO and coconut flakes [8].
- **Desiccated Coconut (DCN) Route**. The DCN route involves washing, grinding, blanching and drying of the coconut meat. The ground meat is dried to a moisture content level of 2.5-3% using a conveyor type hot air dryer with three diminishing temperature levels (100°C, 85°C, and 65°C) (17)°. Desiccated coconut products, which do not pass the accepted quality standards in terms of color and microbial content, can still be processed into high value VCO [8, 17].
- **Grated Coconut Route.** The grated coconut route is similar to the DCN route except that it requires fewer processing steps and equipment [8].

(ii) Fresh-Dry Low-Pressure Extraction Technique

This method is also known as the intermediate moisture content technique [17]. Utilizing a low pressure of approximately 460 psi, oil can be extracted from most seeds and nuts provided the moisture content is maintained between 10-13%. Likewise, oil can be extracted from finely grated and dried coconut meat if moisture content levels are within the optimum range. The grated kernels dried either by indirect hot air drying or solar drying is placed in bags made of cheesecloth [17]. The oil extraction process is carried out in a bridge press, which facilitates easy removal of the residue, and as the cheesecloth acts as a filter, the resultant oil contains only a less amount of fine kernel particles. While the extraction process performed with low moisture content meat results in reduced oil quantity, high moisture content meat yields relatively turbid oil owing to a mixture of oil and coconut milk. The acceptable level of moisture content in VCO should be 0.1% or less [17].

B. Milk Extraction

(i) Modified Kitchen Method

For many decades, people in coconut producing areas like the Philippines and India produced coconut oil for hair and massage applications by boiling the extracted coconut milk from freshly grated or comminuted coconut meat [17]. Nevertheless, the dark yellow oil, thus produced possesses a very short shelf life and becomes sour within three to five days. A slightly modified procedure known as the Modified Kitchen Method utilizes the same principle with the exception that the heating systems are properly controlled so as to prevent the recovered oil from turning yellow. Moreover, the recovered oil is further dried to maintain moisture content less than 0.2%, in order to extend the shelf life. Although, the method produces VCO with an intense coconut aroma, the final recovery of oil is relatively low when compared to other methods due to the fact that a major portion of the oil gets trapped in the protein residue [17].

The modified kitchen method includes two distinct parts, (i) Extraction of coconut milk, and (ii) processing of VCO from the coconut milk. Extraction process can be done manually or through the use of a hydraulic press. The grated kernel is mashed and placed in a cheese bag (manual method) or a white net bag (hydraulic jack method), and tightly squeezed to produce the coconut milk. Often times a second extraction is also performed to enhance the oil quantity. The squeezed milk is then allowed to settle preferably for two hours so as to distinguish the separation of coconut cream (oily phase) and skim milk (aqueous phase) layers [21]. After scooping the coconut cream from the top, the cream is heated maintaining an initial temperature of 90°C for the first hour and later reduced to 80°C to allow the coagulation of protein. And as the oil separates, the temperature is further reduced, and continuous stirring is done to facilitate uniform heat through the mixture. The oil produced is scooped out from the wok carefully and filtered through a cotton-plugged funnel (small scale) or a pressure filter can also be used for large scale processing [17].

(ii) Modified Natural Fermentation Method

The Modified Natural Fermentation Method requires much less investment, with moderate labor and energy inputs compared to all the currently available VCO processing technologies. Dayrit, et al. [22] suggested that the natural fermentation method is a familiar process for the production of VCO which can be conveniently performed both at the household level and at the industrial scale.

This method of VCO production is similar to that of the Modified Kitchen Method except that the extracted coconut milk is diluted and allowed to stand for hours to undergo natural fermentation. Fermentation can also be induced by the addition of foreign substances such as *Lactobacillus plantarum* [23]. The maximum permissible level of FFA content in VCO is 0.1%. However, the FFA content of VCO produced by traditional natural fermentation method ranges from 0.33–0.38%. The reason for such high FFA content in VCO obtained through traditional method could be attributed to the long hours of settling time (36-48 hours). Hence, as a means to obtain water clear VCO with acceptable FFA levels, the fermentation time is regulated to a maximum of 16 hours. The finely grated fresh kernel is subject to a hydraulic press for the extraction of coconut milk. The extracted coconut milk is mixed with one portion of water and left to settle at ~40°C for a maximum of 16 hours for fermentation to occur [24]. The fermenting container displays five distinct layers comprising of different components in ascending order: (i) gummy sediment at the bottom, (ii) fermented skim milk, (iii) fermented curd layer, (iv) VCO layer and (v) a top layer of fermented curd. The top layer of fermented curd is scooped out and the VCO layer is separated and filtered through a filtering funnel plugged with cotton. The fermented skim milk obtained in this process is unfit for human consumption and should be discarded. In certain cases, the coconut milk emulsion can also be separated by adjusting the pH of the emulsion between 3 and 5.6 and inoculated with bacteria cultures [25].

(iii) Centrifuge Method

The best quality VCO can generally be achieved through the centrifuge process [26-27]. There are two different types of centrifuge processing of VCO: (i) the two-phase (liquid-liquid) centrifuge process and (ii) the three phase (liquid-liquid-solid) centrifuge process. Based on how the VCO is recovered from the coconut cream, the two-phase (liquid-liquid) centrifuge process can have slight modifications. Three different routes have been reported. (i) The cream is subjected to vacuum evaporation to remove water and coagulate the protein. (ii) The cream is frozen and heated in a double boiler and is filtered to remove the coagulated protein. (iii) The cream is heated in a controlled temperature to coagulate the protein and remove water. In all the three cases the resultant oil is filtered through a filter press followed by vacuum drying to remove traces of water [28-29].

Compared to the two-phase centrifuge process, the three-phase centrifuge process is simpler: the filtered coconut milk is passed through a three-phase centrifuge system where the individual components are separated by an applied centrifugal force [28]. The coconut milk after its extraction is mixed with one portion of hot water and is fed into the centrifuge. The cloudy oil that comes out of the centrifuge is again mixed with the second portion of hot water and fed again into the centrifuge for a second pass. The final oil is filtered to remove any solid particles and vacuum dried to afford water clear VCO. The scales of operation of VCO are relatively large owing to the high investment cost [17].

C. Coconut Milk Residue

The Bawalan-Masa Process

The Bawalan-Masa Process (BM) is a hybrid of fresh-dry and the fresh-wet processes [17]. The precursor for the VCO extraction in BM process is the coconut milk residue, which represents approximately 25-50% of the weight of the freshly grated coconut meat. The residue is blanched and dried in a mechanical dryer to reach specific moisture content. The defatting of the dried residue under controlled conditions in a specially designed equipment produce VCO and low fat, high fiber coconut flakes. The oil containing fine particles is filtered through a mechanical filter to give a clear VCO. Coconut flakes obtained as a byproduct during the process can be re-dried and ground to produce coconut residue flour. The VCO, thus, produced is very light in texture and is easily absorbed and has a very mild coconut scent [17].

3.2.3. Post Processing Stage

The post processing stage includes all processes that are conducted to improve the quality of the produced VCO. It includes oil drying (removal of moisture content), ageing (removal of the sour smell), and fine filtration (removal of fine residue). Table 3 gives the comparison data of the overall quantity and moisture content of the recovered oil of all the methods discussed so far [5].

3.3. Uses of VCO

VCO has a wide array of uses and applications, which can be classified as either edible or inedible categories. A comprehensive review of its health benefits is discussed in the succeeding sections.

Edible Applications

- VCO serves as an important source of energy in diet [5].
- VCO is used as cooking and frying oil due to its exceptional resistance to rancidity development, and it enhances the flavor of food [5].
- Due to its nature of unchanging palatability, VCO is used as a substitute for buttermilk in filled milk, filled cheese and ice cream [5].

Inedible Applications

- It is used as a skin and hair conditioner [17].
- Aromatherapy and massage oils [30].
- Oil base for a variety of cosmetic and skin care products [31].

Synopsis of Health Benefits

- VCO is the only naturally available low-calorie fat [32].
- VCO boosts the immune system and protects humans from atherosclerosis and cardiovascular disease [32].
- Digestion of VCO takes place easily without the need for bile.
- VCO stimulates metabolism and prevents obesity [33-34].
- It also inhibits cancer causing agents [35].
- It increases the absorption of vitamins, minerals and amino acids [32].
- VCO has the potential to prevent exercise and chronic cold restraint stress-induced damage and restores the antioxidant balance [36].
- The presence of polyphenols and medium-chain fatty acids in VCO imparts anti-stress activity [36].

- Wound-healing rate was increased in skin of rats treated with VCO [19].
- VCO was also used as an 'ethnomedicine' to treat gastrointestinal problems and minor cuts, injuries and swelling [37].
- Effective and safe as mineral oil when used as a moisturizer for mild to moderate xerosis [38].
- Dried-and fermented-processed VCO has hepatoprotective property [39].
- Has anti-oxidant activities and does not adversely affect serum lipid levels [40].
- Has anti-diabetic effects [41].
- VCO displayed inhibition of *Candida sp.* responsible for fungal infection [42].
- The fatty acid present in VCO acts as a potential immunostimulant, which increases immunity through the increase of lymphocyte and Th-CD4 in chickens vaccinated against *Avian influenza* virus [43].

Table 3. Comparative data of VCO obtained by different processes [5, 17]

Type of Process			Quality of Oil	Recovery
Fresh-Dry Processes	High-Pressure Expelling	Wet Milling Route	FFA–0.05-0.08% MC – 0.07-0.1%	60 kg per 100 kg of dried milled kernel
		Desiccated Coconut Route	FFA– 0.05-0.08% MC – 0.0-0.1%	58 kg per 100 kg of desiccated coconut
		Grated Coconut Route	FFA– 0.05-0.08% MC – 0.07-0.1%	30 kg per 100 kg of fresh grated kernel
	Low-Pressure	Low pressure Extraction	FFA– 0.1-0.2% MC – 0.17% & below	25 kg per 100 kg of fresh grated coconut kernel
Fresh-Wet Processes	Coconut Milk Extraction	Modified Kitchen Method	FFA– 0.1% MC – 0.14% & below	16.5 kg per 100 kg of fresh grated coconut kernel
		Modified Natural Fermentation Method	FFA– 0.1% MC – 0.12% & below	34 L per 100 L of coconut milk
		Centrifuge Method (2-phase)	FFA– 0.04-0.08% MC – 0.1% & below	28 L oil per 100 L of coconut milk
	Coconut Milk Residue	Bawalan-Masa Process	FFA– 0.05-0.08% MC – 0.07-0.12%	17 kg per 100 kg of wet residue

4. Worldwide Production of VCO

4.1. How Production of VCO Changed through Time

While it may be the common assumption that the production process of VCO has changed through time, this may not actually be the case. The actual production processes and the methodology behind them have not consistently changed. Fermentation and the unheated and "cold pressed" procedures are the most typical methods used in the mass production of VCO and have not been seen to drastically change over time. These methods are not to be confused with RBD coconut oil (chemically refined, bleached, deodorized) methods [44-45]. However, what has been observed throughout time is an ongoing debate as to which method is actually the best, which is the focus of multiple research groups around the world. For example, a study conducted in 2005 at the University of the Philippines suggested that VCO in general contained more antioxidants than that of RBDs [44]. Nonetheless, the real concern was focused on the question as to which method, fermentation or unheated and "cold pressed," produced more antioxidants. The study concluded that the VCO prepared with heat through the fermentation process contained on the average a significant amount more antioxidants than other methods [44]. A few of these antioxidants included superoxide dismutase, malondialdehyde, Vitamin E, phytosterols, and phenolic compounds, in which were found to have a variety of different functions and benefits [46]. Further studies conducted, which include a study done in 2005 in Malaysia and another in 2011 in Sri Lanka, proved that all methods that involve heating the VCO produces more antioxidants [46-47].

4.2. Largest Producers of VCO

As previously mentioned, the studies conducted were located in a variety of different countries including the Philippines, Malaysia, Sri Lanka, India, etc. The significance of mentioning the location of these studies lies in the fact that these countries are all the major producers of VCO. The leading coconut-producing country is the Philippines, followed by Indonesia, India, Sri Lanka, Thailand, and Malaysia; most of which had universities that conducted significant studies relating to debate between the methods behind producing VCO. There are multiple factors to consider as to why these countries are the ideal location for the production of VCO. Primarily, these countries are located in areas very favorable for coconut palm growth, in which reasons include average rainfall per year, sunlight, sandy soil, and humidity [48-49]. The Philippine coconut industry in and of itself accounts for 1.5% of gross national product (GNP) and is the highest net foreign exchange earner in agricultural exports. The Philippines coconut industry employs nearly 20 million people, which accounts for around one-third of the country's entire population. The industry earns approximately $510 million in U.S. currency on an annual basis. Across the globe, Indonesia and the Philippines were the world's two largest producers of coconuts with an estimated production of 16.3 mn tonnes and 14.4 mn tonnes from 3.3 mn ha and 2.7 mn ha, respectively. India is the third largest with an estimated world coconut production of 10 mn tonnes from 1.9 mn ha. The major exporters were Philippines and Indonesia while India consumed most of its coconut product. Across the Western Hemisphere, Central America, Jamaica, Brazil, and Mexico were the major producers of coconut. The major coconut oil importing countries in the world included EU-15 (15%), the USA (33%), Malaysia (5%), South Korea (4%), East Europe (4%), and Singapore (4%) [50].

Coconut oil accounts for approximately 20% of all vegetable oils used worldwide [51]. Examination of the world coconut oil supply and distribution for over a decade (2002/03 to 2015/16) shows a steady decline in the total supply of the oil starting from 2013/14 (Figure 2). This total supply of the world coconut oil is the sum of the beginning stocks, production, and imports. The decline can be attributed primarily to the decrease in the beginning stocks available for the coconut oil supply. Worldwide exportation of the coconut oil has not tremendously varied over the past five years. Industrial domestic consumption and food use domestic consumption of the coconut oil have also started to increase slightly since 2014/15. Overall, the worldwide ending stocks available for the coconut oil have steadily declined since 2011/12 [52].

4.3. Economic Benefits and Analysis of VCO

VCO was flagged as an unhealthy food product and a possible contributor to heart disease around 40 years ago. However, over the years this idea has been adjusted [53]. Saturated fatty acids in coconut oil gave the oil a bad reputation and resulted in its sparse consumption [53]. However, an increasingly new wave of consumers have flocked to the oil as new studies have proposed health effects including weight loss and analgesic, antipyretic, cardio-protective and anti-carcinogenic effects. The economic boom, which has followed in the wake of these new studies, has generated a very lucrative market in coconut producing countries on a wide scale [54]. In general, VCO production has benefited the local economies, but in other cases the opposite seems to be true.

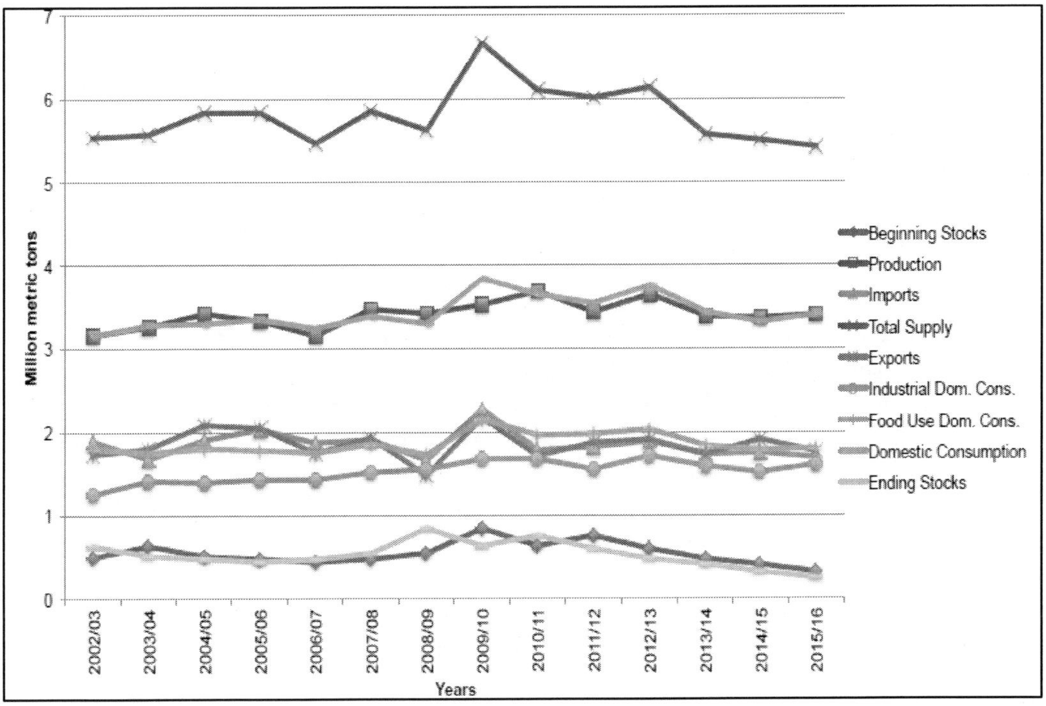

Figure 2. World coconut oil supply and distribution [52].

One of the chief examples of the positive effects is the development of a VCO industry in the Solomon Islands, which before the turn of the century primarily produced copra [55]. After the dried coconut kernel was harvested, it was then shipped off and processed away from the site where it was harvested. The coconut oil demand has brought new prosperity to the islands, which before were plagued with poverty. Dr. Etherington, a Stanford educated economist developed a system called Direct Micro Expelling (DME), which presses the oil on a small scale and obtains it in little under an hour. This allowed the people of the Solomon Islands to expand their use of the coconuts, effectively utilizing more of the coconut at the village level of the coconut oil production [55]. Dr. Etherington's company, Kokonut Pacific, has installed more than 40 DME systems in the Solomon Islands, and sees potential to put in more than 300 in the coming years.

The world's leading producer of coconut oil, the Philippines has shown how lucrative this industry can be. About one third of the population in the Philippines is directly or indirectly employed by the coconut industry in the Philippines [54]. The boom of the coconut oil has generated a massive industry in the Philippines, but some argue that the influx of business has not stimulated the economy at every level. Numerous farmers still live in poverty even though their crop has grown in value. Evidence of this, is based in the facts that coconut farmers are still some of the poorest farmers in the world. It is estimated that sixty percent of the farmers living in the Philippines live in poverty [56].

5. Beneficial Effects of VCO

5.1. Anti-Inflammatory Properties of VCO

Inflammation is a protective response of the immune system against pathogens, but can also result to damaging consequences if not regulated [57]. The complex immune processes and mediators involved in the inflammatory response can induce and aggravate many diseases. Recent studies have found that inflammation plays a key role in various human diseases that are not primarily disorders of the immune system, which includes cancer, atherosclerosis, ischemic heart disease, and neurodegenerative diseases such as Alzheimer's disease [57]. However, current medications have adverse effects and there is a need for more effective and safer anti-inflammatory therapeutics. Thus, VCO offers the potential for further research and development of anti-inflammatory therapies. Different groups have shown great interest in exploring the anti-inflammatory effects of VCO because of its high phenolic contents. Some of the phenolic acids identified in VCO include caffeic acid, p-coumaric acid and feulic acid [58]. Polyphenols control and reduce inflammation through a series of pathways, therefore, preventing cancer and other diseases with an inflammatory pathogenesis. Therefore, it is of interest of this section to summarize the anti-inflammatory action of VCO that has been investigated [59].

5.1.1. Anti-Inflammatory and Anti-Oxidant Properties of VCO Inanimal Models of Rheumatoid Arthritis

Arthritis is caused by joint inflammation that can ultimately lead to disability. Rheumatoid Arthritis (RA) in particular is a chronic inflammation that affects synovial joints and progressively results to destruction of articular cartilage. Current medications used to treat RA have side effects; the need for effective therapeutics that also permit greater compliance is what drive the investigation into products such as VCO. Vysakh et al. explored the anti-inflammatory effect of the polyphenolic fraction extracted from VCO on experimental arthritis [60]. Their results demonstrated the potential health benefit of VCO on adjuvant induced arthritis in rats, and determined that mechanism behind this action is due to its antioxidant and anti-inflammatory effects. Their findings on how VCO inhibits arthritis can be summarized as follows:

- **Downregulated (inducible nitric oxide (NO) synthase) iNOS expression.** Nitrite production in VCO treated animals was reduced significantly compared to adjuvant treated rat. This suggests that reduction of chronic inflammation in arthritic animals might be due to the downregulation of iNOS expression followed by decreased cellular production of NO. Therefore, the anti-inflammatory activity of VCO may be due to their ability to counteract NO induced oxidative damage, which eventually helps in the remodeling of cells.

- **Decreased cyclooxygenase-2 (COX-2) expressions.** During inflammation, COX-2 enhances the synthesis of prostaglandins (PGs) that are associated with inflammation. The activity of total COX and COX-2 expressions was decreased in VCO treated rats. Reduction of paw swelling and decreased expression of COX-2 indicate the immunological protection provided by VCO.
- **Inhibited tumor necrosis factors-α (TNF-α) activity.** TNF-α may play a role in the pathology of RA. Blocking of TNF-α followed by inactivation of T cells, macrophages, and fibroblasts interrupts the inflammatory process. This anti-TNF-α activity supports further pharmacological research on VCO, in hopes of improving anti-inflammatory medications.
- **Downregulated Interleukin-6 (IL-6) expression.** IL-6 plays a devastating role in cartilage and bone degradation during arthritis. VCO supplementation significantly downregulated the expression of IL-6 in paw tissue of arthritic rats. VCO supplementation also restored the levels of acute-phase proteins (APPs) to near normal values in arthritic rats. Thus, decreased IL-6 production correlated with decreased APP production, thereby reducing disease severity.
- **Enhanced Glutathione (GSH) and superoxide dismutase (SOD) activity**. Polyphenols from VCO enhanced the activity of GSH and SOD, which help preserve the integrity of cellular membranes. VCO administration produced a significant increase in free radical scavenging activity of antioxidant enzymes like glutathione peroxidase (GPX) and catalase.
- **Decreased lipid peroxidation.** The elevated levels of lipid peroxidation were significantly decreased after VCO treatment. VCO reduces free radicals formation as well as inflammation. These results imply that adjuvant induced arthritis may be associated with lipid peroxidation and demonstrates the anti-arthritic effect of VCO. VCO reduced lipid peroxidation causing a modulation in cellular antioxidant defense system.

Intahphuak et al. also explored the inhibitory activity of VCO against acute and chronic inflammation. They found that VCO exhibited moderate anti-inflammation on carageenin- and arachidonic acid (AA)-induced hind paw edema in rats as well as on ethyl phenylpropiolate-induced ear edema, which are all models of acute inflammation [59]. Consistent with the findings by Vysakh et al., they also demonstrated that the consumption of VCO aided in inhibiting the expected immune factor responses to endotoxin, and diminished the production of pro-inflammatory cytokines *in vivo*. They also found that the mode of action of VCO to inhibit inflammation may also involve the lipoxygenase pathway because it is effective in AA-induced rat paw edema model. AA-induced edema animal model has been reported to be resistant to selective COX inhibitors but sensitive for detecting the *in vivo* anti-inflammatory activity of lipoxygenase inhibitors [61].

The group also investigated the effects of VCO on cotton pellet-induced granuloma formation in rats and compared it to the effects of the known anti-inflammatory drugs indomethacin, and prednisolone. Granulomatous inflammation is a distinctive pattern of chronic inflammatory reaction. Indomethacin and prednisolone inhibited granuloma formation by interfering with the proliferative phase of inflammation. However, both drugs also caused weight gain. They have shown that VCO also inhibited granuloma formation but did not result in weight gain. VCO exhibited an inhibitory effect on chronic inflammation by reducing the transudative weight, granuloma formation, and serumalkaline phosphatase activity. The results obtained suggest that VCO probably inhibits the proliferative phase of chronic inflammation but avoids the steroidal-like effect, as it had no effect on body weight gain and dry thymus weight. The activity of VCO on the alkaline phosphatase level in serum may be due to its inhibitory effect on inflammatory cell activity and/or stabilization of the lysosomal membrane.

Similar to the paper of Intahphuak et al. [59], Zakaria et al. [85] investigated the anti-inflammatory effects of VCO using *in vivo* models. The group utilized two types of VCO, produced by standard drying (VCOA) and fermentation (VCOB) on animal models (mice/rats) of both acute and chronic inflammation. The VCOs exhibited anti-inflammatory activity in an acute model (carrageenan-induced paw edema test), consistent with the findings of Intahphuak et al. with the same animal model. Comparing the two, VCOB showed a greater anti-inflammatory effect than VCOA. However, in contrast to the findings of Intahpuak et al. [59], the VCOs did not show ant-inflammatory effects on chronic (cotton-pellet-induced granuloma test) model of inflammation.

5.1.2. Anti-Inflammatory Effect of VCO against E.Coli Endotoxin

Dietary fat influences many aspects of immune function. *Escherichia coli* endotoxin is a potent stimulator of interleukin 1 production from macrophages. The present study examines how feeding rats high-fat diets either rich (corn oil) or poor (coconut oil) in linoleate at high and low concentrations affect responses to endotoxin. Spleen phosphatidylcholine linoleate contents were higher in the corn oil than in the coconut oil group and arachidonate concentrations were highest in the group fed a high concentration of corn oil. Coconut oil completely abolished the responses to endotoxin. The inhibitory effects of coconut oil against the endotoxin largely be due to reduced prostaglandin and leukotriene synthesis [62].

Similar to the aforementioned study, Sadeghi et al. (1999) carried out an investigation on the effect of dietary oils such as coconut oil, corn oil, olive oil, safflower oil and fish oil on *in vivo* cytokine response to bacterial lipopolysaccharide [63]. These dietary oils were fed to mice for 5 weeks and the mice were then injected with a non-lethal dose of *Escherichia coli.* The mice were sacrificed after 90 or 180 minutes post *E. Coli* injection to measure plasma cytikine concntrations. The results showed lower peak plasma concentrations of TNF-alpha, IL-1beta and IL-6 in mice fed with coconut oil and fish oil compared to other dietary oils. Moreover, peak plasma IL-10 concentrations were higher in mice fed with coconut oil than in mice fed with other oils. These results suggest that coconut oil diminishes production of pro-inflammatory cytokines *in vivo* and diminishes enhanced production of IL-10. This appears to be an additional anti-inflammatory effect of this oil, which could give added benefit in various clinical conditions.

5.1.3. Therapeutic Effect of VCO on Inflammatory Acne Vulgaris

VCO contains a large amount of lauric acid, which has been shown to have antibacterial property on *Propionibacterium acnes* (*P. acnes*). This served as the motivation to investigate the anti-inflammatory property of lauric acid on acnes vulgaris. The anti-inflammatory activity of VCO has also been explored on Acne vulgaris. Acne vulgaris or simply acne, is the most common human skin infection and its prevalence is about 80% in the majority of the countries worldwide [64]. Inflammatory lesions of acne may lead to acne scarring and may have an impact on psychosocial health [65]. Acne is caused by *P. acnes*, a Gram-positive anaerobic bacterium that dwells in the pilosebaceous follicles of the skin. Although *P. acnes* is part of the normal skin bacterial flora, it plays a key role in the development of inflammatory acne when it proliferates and colonizes the pilosebaceous unit. Previous studies have found that *P. acnes* triggers the production of proinflammatory cytokines that are mediated by Toll-like receptors [66]. In another study, Nakatsuji et al. demonstrated the potential of lauric acid as an alternative option for antibacterial therapy in acne treatment. Lauric acid showed stronger antimicrobial activity as compared with benzoyl peroxide against skin bacteria, including *P. acnes, in vitro*. It also showed therapeutic potential against *P. acnes*-induced inflammation *in vivo*. This study evaluated the antimicrobial property of lauric acid against *P. acnes* both *in vitro* and *in vivo*. Both intradermal injection and epicutaneous application of lauric acid effectively decreased the number of *P. acnes* colonized with mouse ears, thereby relieving *P. acnes*-induced ear swelling and granulomatous inflammation. This promising result showed the potential of using lauric acid as an alternative treatment for antibiotic therapy of *acne vulgaris* [67].

5.2. Analgesic and Antipyretic Potential of VCO

Intahphuak et al. found that VCO demonstrated a moderate analgesic effect on acetic acid-induced writhing response in mice [59]. VCO at high doses (1000-4000 mg/kg) decreased the number of writhes induced by acetic acid significantly (~30-60%). This effect is modest compared to the known non-steroidal anti-inflammatory drug (NSAID) drug, Indomethacin, which only requires 10 mg/kg dose to markedly decrease the writhing response by ~80%. The mechanism of acetic acid-induced algesia involves the release of endogenous substances such as H^+, K^+, serotonin, histamine, bradykinin, PGs, and substance that excite pain-nerve endings. The analgesic effect of VCO could be due to its ability to block both the synthesis and release of these endogenous substances responsible for pain.

The analgesic effect of VCO was also assessed by Zakaria et al. [85] using several tests. The group utilized acetic-acid induced abdominal constrictions, hot plate test and formalin-induced paw licking test on mice or rats. The VCOs (VCOA and VCOB) exhibited analgesic effect in both the abdominal constriction test and hot plate test indicating their activity in blocking both peripherally and centrally mediated pain induced by chemical and thermal stimuli. The peripheral and central effects of VCOs were further confirmed by their effectiveness in reducing pain in the formalin test. Compared to the analgesics, acetyl acetic acid and morphine, VCOs were less effective even at higher concentration.

5.2.1. Antipyretic Effect Effects of VCO on Yeast-Induced Hyperthermia in Rats

Intahphuak et al. also studied the antipyretic potential of VCO. Fever is a clinical sign of inflammation; elevation in body temperature occurs when the concentration of PGE2 in certain parts of the brain increase. The group demonstrated that VCO exhibited antipyretic activity on yeast-induced hyperthermia. The antipyretic activity of VCO most probably is due to the inhibition of cyclooxygenase, therefore blocking the synthesis or release of PGs in the thermoregulatory center. This mode of action is similar to that of Ibuprofen, a non-steroidal anti-inflammatory agent that possesses antipyretic activity [59].

5.3. Antioxidant, Anti-Stress, and Anticancer Benefits

Mainstream popularity of VCO has led to misinformation regarding potential benefits and how it is beneficial. For example, many companies touted the Vitamin E content of VCO and interchange the benefits of the two. Vitamin E comes in several forms and refers to a group of compounds that include tocopherols and tocotrienols. Coconuts have been shown to have modest tocopherol and tocotrienol concentrations of 0.07, 0.79, 0.18, and 1.04 mg/100 g edible weight of γ-T, α-T3, γ-T3, and total, respectively [68]. Regarding potential anticancer properties, Jordan, et al. developed tocopherol-rich nanoemulsions which were shown to possess anticancer activity [69]. This finding shows that VCO may possess some anticancer activity due to the presence of tocopherols and tocotrienols. Also, in a separate study, VCO enriched with zinc was shown to increase the number of helper T_{cd4} and cytotoxic T_c cells [70]. T_c cells release cytokines, which are capable of activating macrophages, thereby eliminating virus-infected cells and destroying cancer cells. Thus, VCO appears to have some anticancer value. However there are a couple of things to bear in mind. One is that VCO only has a small amount of vitamin E and many other fruits and vegetables have more of these tocopherols and tocotrienols. Another thing to note is that to the author's knowledge there appears limited evidence as of yet to support VCO having significant anticancer properties. The greater body of working comparison, instead supports VCO having antioxidant, anti-inflammatory, and immunomodulatory effects [71].

In the body, lipoproteins carry cholesterol through the body, to or away from cells. Low-density lipoprotein (LDL) carries cholesterol through the body and into cells. High-density lipoprotein (HDL) carries cholesterol away from cells. When LDL is carrying cholesterol into the cells of arteries, it can become trapped within artery walls. If there are free radicals within the body or other oxidizing substances, plaque can form and restrict blood vessels [72].

The lipid peroxides are typically unstable, decomposing to form a complex series of compounds [69, 71, 73-74]. These newly formed compounds often include reactive carbonyl compounds comprised of reactive carbon-oxygen double bonds. Malondialdehyde (MDA) is a product formed from the decomposition of polyunsaturated fatty acid peroxides. Measurement of the MDA is an indicator of lipid peroxidation. The assays often used include chromogenic reagent, which reacts with MDA to form a stable chromophore whose absorbance maximum can then be measured by UV-VIS spectrophotometry. Antioxidant activity is assessed by measuring lipid peroxidation/oxidative stress and antioxidant capacity through the evaluation of the MDA concentration and other antioxidant enzyme levels, which often increase in response [69, 71, 73-74]. These enzymes may include catalase, glutathione peroxidase (GPX), glutathione reductase, and superoxide dismutase (SOD). The presence of polyphenols and the medium chain fatty acids are said to contribute to the antioxidant activity/antioxidant capacity of VCO.

Lipid peroxidation from free radicals and oxidants is a well-established injury mechanism in both plants and animals and is used to indicate oxidative stress in cells as well as tissues [69, 71, 73-74]. Oxidative stress can occur due to this free radical generation in the body leading to bone stress as well as organ damage, including the heart. Antioxidants generally do not reduce cholesterol levels, but rather they prevent plaque from building up by binding to and removing free radical scavengers and other oxidants, thus preventing LDL oxidation. The polyphenol groups present in VCO were found to be capable of preventing *in vitro* LDL oxidation, thus, VCO does provide antioxidant benefits and numerous studies have shown these benefits when VCO has been added to the rat diet [69, 73-77]. In one of these studies, VCO had a greater inhibitory effect on microsomal lipid peroxidation compared to copra and groundnut oil and so is a better antioxidant [73]. The VCO increased the level of antioxidant enzymes and, thus, was able to prevent lipid peroxidation. Findings in this study showed that VCO also reduced LDL and increased HDL cholesterol and total cholesterol levels [73].

The antioxidant properties of coconut oil obtained through the various methods have often been examined. The antioxidant properties of coconut oil obtained from different extraction methods are different in scale. For example, Seneviratne, et al. concluded that a hot method of extraction of coconut oil can produce an oil that contains more phenolic compounds than a coconut oil extracted under cold conditions [78]. As a result of this, coconut oil extracted under hot conditions showed higher antioxidant potential than coconut oil extracted under cold conditions since the phenolic compounds are the free radical scavengers. Therefore, consumption of VCO that have undergone different methods of processing may result in VCO with different antioxidant content and, thus, slightly different potential health benefits. Other extraction methods for coconut oil have been examined as well with a goal of using only as little processing as can be to keep the properties of the virgin material. An extraction using supercritical carbon dioxide resulted in 99% extraction efficiency [79]. Use of this method would result in minimal alteration of virgin material and maximum recovery of oil, however it would have a high extraction cost.

The properties of coconut oil may vary in look, taste and composition depending on methods used to obtain the oil from the nut, regardless of whether the method involves processing (non-VCO) or not (VCO) (3-4, 46). For example, in one study, VCO was obtained through two separate methods: through chilling and also through fermentation, and both were compared to each other as well as coconut oil obtained through the RBD process [80]. The VCO obtained through fermentation showed a stronger scavenging effect and higher antioxidant activity than the VCO obtained from the chilling method. The VCO from use of the chilling method showed a higher reducing power than fermentation obtained VCO and the RBD oil. The VCO obtained via either method showed a greater antioxidant capacity and higher phenolic acid content than the RBD processed oil.

The loss/change in hormone such as estrogen through naturally growing older or other means leads to a progressive accumulation of oxidative damage in tissue and bone [81]. In 2012, Abujazia, et al., showed that addition of 8% VCO (8 g VCO to 100 g chow) to the rat diet caused a significant decrease in stress-induced MDA levels in bone [81]. In these studies, female rats were subjected to ovariectomy or sham manipulation in order to simulate postmenopausal osteoporosis. The removal or manipulation of the ovaries and subsequent change in hormones leads to progressive loss of bone matrix in the rat, and, thus, a useful model for the occurrence in postmenopausal women. Osteoporosis is bone inflammation associated with oxidative stress and in this study the stress due to lipid peroxidation was estimated by an MDA assay kit using tibia bone sample. Rats subjected to ovariectomy also had higher GPX and SOD concentrations compared to control groups which seems to indicate that VCO prevents the lipid peroxidation and increases the concentrations of antioxidant enzymes in the rat model used. Although the SOD concentration was increased, the increase was not statistically significant unlike the MDA and GPX concentrations, which were statistically significant and the results of these studies indicate alignment with other studies showing the effectiveness of VCO in maintaining bone structure. VCO prevented loss of bone density by preserving bone mass in the rat model.

A similar study from the same group and also using the postmenopausal osteoporosis rat model was performed by Hayatullina, et al. using an 8% (8 g VCO to 100 g chow) VCO supplemented diet [82]. A diet supplemented with VCO led to rats with significantly greater bone volume in rats that had ovariectomy or sham manipulation than those without VCO in the diet. These findings support the evidence for reduction of bone loss density through lipid peroxidation in the estrogen deficient ovariectomized rat by inclusion of a VCO enhanced diet.

VCO then, clearly is effective as an antioxidant when included in the diet. How might VCO compare as an antioxidant to several other oils deemed healthy enough to be included as part of a balanced diet? Findings by Arunima, et al. revealed that a diet supplemented with VCO was more effective at reducing oxidative stress in rats than olive, sunflower and copra oil [36, 83]. In these studies, VCO functioned to help increase several enzyme activities including those of catalase, glutathione peroxidase, and glutathione reductase and superoxide dismutase. Enzymatic activity in the rat model used in this work supports findings indicating the antioxidant ability of VCO as part of a balanced diet. VCO also decreased tissue lipid levels in this study as it did in those previously mentioned.

Studies in similar animal models regarding the antioxidant capacity of VCO have continued to reinforce findings from previous rat studies. Yeap examined *in vivo* antioxidant and anti-stress properties of VCO in mice [36]. The studies showed that stress induced lipid peroxidation from mice administered the forced swim test and cold restraint stress test at 4°C was reduced in mice serum through the use of VCO at doses of 10 ml/kg animal weight. VCO also caused an increase in the liver superoxide dismutase enzyme level and administration led to restoration of balance in brain monoamine neurotransmitter levels, particularly serotonin. Since administration of VCO blocked the increase of serotonin seen in untreated, stressed animals, the depletion of this neurotransmitter was not observed in the VCO-treated animals. This study demonstrated the potential for VCO in blocking stress-induced inflammation and lipid peroxidation.

Thus, VCO consumption as part of a balanced diet is beneficial due to the known antioxidant/antistress properties. It turns out that antioxidant properties are also seen in the topical application of VCO in addition to dietary intake of VCO. Nevin, et al., 2010, showed that rats treated topically with VCO to excision wounds healed faster. 0.5 ml and 1.0 ml of VCO was applied to wounded Sprague-Dawley rats [19]. Lipid peroxide levels were lower in the VCO treated wounds versus control. Antioxidant activity was also proven based on the alteration of enzyme levels of glutathione and MDA. Thus topical application of VCO is beneficial in addition to dietary intake.

Clearly, we have seen how VCO can reduce lipid peroxidation and alter enzyme levels associated with oxidative stress. VCO can eliminate or reduce the loss in bone density due to accumulation of oxidative damage. VCO can also reduce or eliminate oxidative damage in body organs as well. For example, oxidative stress can lead to hepatic toxicity [84]. Treatment with VCO was shown to help protect liver function, reduce liver damage associated with lipid peroxidation and lead to an improvement in hepatic antioxidant enzymes, enzyme activity and liver fatty acid level [39, 85-86]. Treatment with VCO was shown to help protect liver function, reduce liver damage associated with lipid peroxidation and lead to an improvement in hepatic antioxidant enzymes, enzyme activity and liver fatty acid level

5.4. Cardioprotective Effects

VCO has a rich content of MCFAs consisting of caproic acid, caprylic acid, capric acid, and lauric acid [10]. It contains high amount (65%) of medium chain triglycerides (MCTs). These MCTs are directly absorbed from the intestinal tract and sent directly to the liver and doesn't participate in the biosynthesis and transport of cholesterol [87] and, thereby provides a quick source of energy. As a result, VCO was found to be effective against cholesterol levels. VCO was found to reduce the total cholesterol, triglyceride, phospholipid, and LDL, and increase the HDL in the serum and tissues [73]. In addition, the polyphenolic component found in VCO was capable of reducing lipid levels and LDL significantly [74].

VCO was also reported to prevent hypertension and improves endothelial functions in rats fed with repeatedly heated palm oil [88]. In another similar study, it was reported that VCO supplementation demonstrated a cardioprotective effect by preventing an elevated blood pressure when rats were fed with repeatedly heated palm oil [89]. VCO at a dose of 1.43 ml/kg reported protective effects on the vascular and cardiac tissue remodeling when rats were fed with repeatedly heated palm oil [90].

The higher reaction polyphenols in VCO are responsible for its anti-inflammatory and anti-oxidant effects and therefore helps in the prevention of cardiovascular disease and atherosclerosis [60]. Animals fed with VCO were found to have better coagulation studies with lower fibrin levels and better prothrombin time when compared with copra oil and sunflower oil [12]. VCO administration to rats increased anti-oxidant activity and lowered lipids and thrombotic factors compared to normal coconut oil [76].

In a prospective open label trial in humans, it was found that a 4-week supplementation of VCO significantly reduced waist circumference and improved lipid profile and it is safe for use in humans [91]. In another study, it was found that coconut oil consumption did not elevate serum total cholesterol or serum triglycerides in a cohort of 1,839 Filipino women [92].

As described in this section, the existing literature reports the cardiovascular health benefits of VCO. However, evidence pertaining to VCO use is limited and more research in this area is needed to definitively recommend dietary VCO to improve cardiovascular disease risk.

5.5. Bone Loss Prevention

Oxidative stress and free radicals are implicated in the pathogenesis of osteoporosis. Therefore, antioxidants are likely to prevent the disease. In one study, it was shown that VCO effectively improved bone structure and prevented bone loss in osteoporosis rats and this effect can be attributed to the polyphenols present in VCO [82]. Further, VCO supplementation showed a significant improvement in the bone antioxidant status by preventing lipid peroxidation and increasing levels of glutathione peroxidase and superoxide dismutase enzymes in the osteoporotic rat model [81].

5.6. Antimicrobial Effects

VCO has a long history of use as an antibacterial agent. A history of safe topical use and no known or reported cases of adverse effects opens up more possibilities of the use of VCO against infections. VCO contains high quantities of MCFA like lauric acid, caproic acid, and caprylic acid; studies have shown that these MCFA are responsible for its antibacterial, antifungal, antiviral, and properties [93].

- ***Antibacterial activity***. MCFAs and their derivatives are effective in destroying lipid-coated bacteria by disintegrating their lipid membrane. The antimicrobial activity of VCO might be due to an active compound monolaurin, which is a product of lauric acid metabolization [94]. Lauric acid is the predominant fatty acid in the coconut oil [95]. It is also present in breast milk and was found to helpsupport healthy growth in breastfed infants and was shown to possess antimicrobial properties [96].
  VCO and monolaurin have shown antibacterial effects on *Staphylococcus aureus*, and can be useful in the proactive treatment of atopic dermatitis colonization [97]. Similarly, in pediatric patients with mild to moderate atopic dermatitis, a topical application of VCO for 8 weeks was superior to that of mineral oil based on clinical and instrumental assessments [98]. In contrast, VCO did not inhibit the growth of *Staphylococcus aureus* and the morphology of *Staphylococcus aureus* cells exposed to the oil was not different from that of the untreated cells. This effect might be attributed to the low concentration of lauric acid (0.47 mg/ml) in VCO, which is below the minimum inhibitory concentration (MIC) of lauric acid (1.6 mg/ml) [99]. Another study demonstrated the inhibition of growth of antibiotic resistant *Clostridium difficile* mediated by MCFAs, which are derived from VCO [100]. Hydrolyzed VCO was more effective against *Pseudomonas aeruginosa*, while unhydrolyzed VCO did not inhibit the bacterial growth [101]. In another study, enzymatic hydrolysis of VCO inhibited growth of Salmonella species in *in vitro* and *in vivo* studies [102].
- ***Antifungal activity***. Since VCO is a rich source of MCFA, which possesses antifungal activity, a study in Nigeria reported its effectiveness as an antifungal agent, and compared its action to fluconazole, a first line of treatment against drug resistant *Candidaalbicans*. The study concluded the oil to be very potent against *Candida species* at 100% concentration when compared to fluconazole and therefore, can be used in the treatment of fungal infections caused by *Candidaspecies* [42].

Capric acid was more effective against *Candidaalbicans*, while lauric acid was the most active at lower concentrations.

- ***Antiviral activity***. VCO was found to be effective against lipid-coated viruses, such as Epstein-Barr Virus (EBV), influenza virus, leukemia virus, hepatitis C virus, and Cytomegalovirus (CMV); it acts by disrupting viral membranes, assembly, and maturation. Lauric acid has greater antiviral activity than caprylic acid, capric acid, or myristic acid [103]. Monolaurin acts by solubilizing the lipids and phospholipids in the envelope of the virus and, thereby causes disintegration of the envelope [53].

CONCLUSION

VCO is becoming popular as a functional food oil due to increasing public awareness about its health benefits. VCO is obtained from the fresh, mature kernel (meat) of the coconut by mechanical or natural means, with or without the use of heat, without undergoing chemical refining, bleaching or de-odorizing process, which does not lead to the alteration of the nature of the oil and preserves the essential amino acids, tocopherols (vitamin E) and other valuable compounds present in coconut oil. Recently published literature about the health benefits of VCO has been a boost for the growing demand of VCO.

VCO has a rich content of MCFAs, predominantly lauric acid and some SFAs with a higher amount of myristic acid compared to other SFAs. The properties such as higher levels of saponification, phenolic compounds and tocopherols and low iodine numbers and peroxide values make VCO a potential healthy addition to the normal diet.

As discussed before, there are several methods by which one can produce VCO. Accordingly, the quality and quantity of the final oil produced vary with the different processing techniques employed for extraction. Also, the produced VCO possess varied organoleptic characteristics. The high-pressure expelling method and Bawalan-Masa method generally produce VCO with a longer shelf life of one year or more. The modified kitchen method incurs a very low investment cost and is prone to rancidity development after five days if moisture is not properly removed after extraction. A two-stage centrifuge process can produce best quality VCO with coconut aroma. However, the best oil recovery is achieved with a high pressure expelling method. Hence, in order to ensure that only high quality VCO is produced, it is highly recommended adhering to the principles of good manufacturing practices.

Different groups in animal models have evaluated the anti-inflammatory, analgesic and antipyretic properties of VCO. VCO inhibits both acute and chronic inflammation in RA induced animal models when used in high dose. In other studies, VCO also exhibited moderate analgesic and antipyretic effect on acetic acid-induced pain and yeast-induced hyperthermia, respectively. There are very few studies on the analgesic and antipyretic activity of VCO and, thus, needs further investigation. These preliminary studies give encouraging results to further exploit the therapeutic potential of VCO.

On the other hand, other studies proved that there are indeed beneficial antioxidant properties in VCO. A diet enhanced with VCO may have potential health benefits as long as the fatty acid content in the diet is moderated. Unadulterated VCO may also be richer in antioxidants providing greater health benefits than RBD or other means of obtaining the coconut oil. Of course, maintaining a proper balance of saturated fatty acids in the diet is important and this should be kept in mind when VCO is included in dietary intake.

VCO with limited pharmacotherapeutic properties is gaining popularity in the modern society. From the existing literature, there are many health benefits including cardioprotective effects, bone loss prevention, antibacterial, antifungal, and antiviral effects. However, more research is needed to provide conclusive evidence against clinical applications.

FUTURE DIRECTIONS

VCO is currently enjoying a great deal of positive press thanks to increased publicity about its potential health benefits. Although it is being touted as the latest "superfood," with ascribed qualities that run the gamut from causing weight loss to fighting cancer, it is likely that future research on VCO will focus on its anti-inflammatory properties. This is because inflammation is now attracting attention from scientists as a potential unifying factor among multiple diseases. Recent studies have shown that tissue degeneration due to chronic inflammation plays a role in the pathogenesis of such different diseases as type 2 diabetes, Alzheimer's, and age-related macular degeneration [105]. While inflammatory mediators are usually produced in response to cellular injury, they are also released from fat cells, which explains the increased risk that overweight people face for a variety of conditions.

A 2016 study published in the neurology journal *Brain* showed that brain inflammation is likely a driving factor behind Alzheimer's disease, rather than simply an immune reaction after brain pathology has already set in [106] Treatment of inflammation in Alzheimer's mice stalled their loss of neuronal connections, and improved their memory loss and behavioral problems. However, treatment did not stop the progression of amyloid plaque buildup in the mice's brains. Interestingly, the results of this study dovetail with those of a 2015 study published by the University of Valencia, which found that administration of 40 mL/day of VCO to Alzheimer's patients produced a statistically significant improvement in cognitive status [107]. It is important to note, however, that this study ascribed the beneficial affects of VCO to the cellular energy source, namely ketones, provided by its medium-chain triglycerides. Other studies with similar results also prefer the ketogenic hypothesis to the anti-inflammatory hypothesis [108]. Nevertheless, considering the novelty of the association between Alzheimer's disease and inflammation, it is likely that more light will be shed on how much of a role VCO's anti-inflammatory properties play in its therapeutic effects on Alzheimer's patients.

The ketogenic and oxidizing properties of VCO have also attracted attention in recent years, particularly in the context of weight loss. The most recent example is the craze for so-called "bulletproof coffee," which was claimed to heighten alertness, suppress appetite, and increase fat burning via ketosis, due to the addition of butter and medium-chain triglycerides (often in the form of VCO). The current scientific standpoint towards bulletproof coffee is one of skepticism, considering the variability of results and lack of formal research. However, the potential association between weight loss/ketogenesis and VCO alone is a current hot topic in research. A study due out in 2016 correlates a high-VCO diet with improvement in hyperglycemia and dyslipidemia in high fructose-fed rats [109]. A 2014 study reported that

VCO significantly increased hepatic lipid metabolism and fatty acid oxidation compared to copra oil [110]. Research in this area is ongoing.

In summary, the future directions of VCO research seem to focus most on its anti-inflammatory and ketogenic/lipid oxidizing properties. Furthermore, both these areas of research appear most popular in the context of treating Alzheimer's disease and promoting weight loss. While results for these topics of investigation remain preliminary, they are the focus of nearly all the most recent papers published on the health effects of VCO. We may, thus, anticipate that research into the correlation of VCO administration with inflammation, lipid levels, weight loss, and Alzheimer's is ongoing, and that more complete information is forthcoming.

REFERENCES

[1] Guarte, RC; Mühlbauer, W; Kellert, M. Drying characteristics of copra and quality of copra and coconut oil. *Postharvest Biol Technol* [Internet]. 1996 Dec [cited 2016 Jan 26], 9(3), 361–72. Available from: http://www.sciencedirect.com/science/article/pii/S0925521496000324.

[2] Fife, B. Virgin Coconut Oil: Nature's Miracle Medicine. Piccadilly Books, Ltd., 2006. 100 p.

[3] Marina, AM; Man, YBC; Nazimah, SaH; Amin, I. Chemical Properties of Virgin Coconut Oil. *J Am Oil Chem Soc [Internet].*, 2009 Jan 24 [cited 2016 Jan 23], 86(4), 301–7. Available from: http://l ink.springer.com/article/10.1007/s11746-009-1351-1.

[4] Marina, AM; Che Man, YB; Amin, I. Virgin coconut oil: emerging functional food oil. *Trends Food Sci Technol [Internet].*, 2009 Oct [cited 2016 Jan 23], 20(10), 481–7. Available from: http://www.sciencedirect.com/science/article/pii/S0924224409002052.

[5] Bawalan, DD; Chapman, KR. *Virgin coconut oil production manual for micro-and village-scale processing.*, 2006 [cited 2016 Jan 24], Available from: http://agris.fao.org/agris-search/search.do?recordID= XF2006 427761.

[6] Adkins, SW; Foale, M; Samosir, YMS. Coconut revival: new possibilities for the "tree of life." *In Australian Centre for International Agricultural Research*, 2006 [cited 2016 Jan 26]. Available from: http://espace.library.uq.edu.au/view/UQ:108025.

[7] dela Cruz, LD. Potential of Virgin Coconut Oil in the Production of Lacquer Enamel Paint. *In: Proceedings of the World Congress on Engineering and Computer Science.*, 2010.

[8] Peter, KV; Alice, K; Bavappa, KVA. Commercial Crops Technology (Horticulture Science Series-8). *New India Pub. Agency*, 2007.

[9] Philippine Coconut Authority. 2013 Outlook for the Coconut Industry. *Bangko Sentral ng Pilipinas*, 2013 Mar.

[10] Mansor, T; CheMan, Y; Shuhaimi, M; Abdul-Afiq, M; Ku-Nurul, F. Physicochemical properties of virgin coconut oil extracted from different processing methods. *Int Food Res J.*, 2012, 19(3), 837–45.

[11] Man, YBC; Karim, MIBA; Teng, CT. Extraction of coconut oil with Lactobacillus plantarum 1041 IAM. *J Am Oil Chem Soc [Internet].*, 1997 Sep [cited 2016 Jan 23],

74(9), 1115–9. Available from: http://link.springer.com/article/10.1007/s11746-997-0033-0.

[12] Babu, AS; Veluswamy, SK; Arena, R; Guazzi, M; Lavie, CJ. Virgin coconut oil and its potential cardioprotective effects. *Postgrad Med.*, 2014 Nov, 126(7), 76–83.

[13] Arlee, R; Suanphairoch, S; Pakdeechanuan, P. Differences in chemical components and antioxidant-related substances in virgin coconut oil from coconut hybrids and their parents. *Int Food Res J.*, 2013, 20(5), 2103–9.

[14] SECTION 2. Codex Standards for Fats and Oils from Vegetable Sources [Internet]. [cited 2016 Jan 23]. Available from: http://www.fao.org/docrep/004/y2774e/y2774e04.htm.

[15] Bureau of Product Standards. Philippine National Standard: Virgin coconut oil. Department of Trade and Industry Philippines, 2004.

[16] O'brien, RD. Fats and oils: formulating and processing for applications [Internet]. CRC press, 2008 [cited 2016 Jan 24]. Available from: https://books.google.com/books?hl=en&lr=&id=3wpHj3mvra8C&oi=fnd&pg=PP1&dq=Fats+and+oils:+Formulating+and+processing+for+applications&ots=c4FC0qL9zc&sig=5CkiD-MPqXwHkkChc9hi WdQw1Pk.

[17] Bawalan, D. Processing Manual for Virgin Coconut Oil, its Products and By-products for Pacific Island Countries and Territories. New Caledonia Secr Pac Community. 2011.

[18] Prasangika, JPC; Jayasundera, J; Asanka, JRK; Marikkar, JMN. Production of dried pulverized kernel for virgin coconut oil extraction: Assessment on particle size distribution, *drying curve pattern and quality characteristics.*, 2008 [cited 2016 Jan 29], Available from: http://cri.nsf.ac.lk/handle/1/3857.

[19] Nevin, KG; Rajamohan, T. Effect of Topical Application of Virgin Coconut Oil on Skin Components and Antioxidant Status during Dermal Wound Healing in Young Rats. *Skin Pharmacol Physiol [Internet]*, 2010 [cited 2016 Jan 24], 23(6), 290–7. Available from: http://www. karger.com/doi/10.1159/000313516.

[20] Raghavendra, SN; Raghavarao, KSMS. Effect of different treatments for the destabilization of coconut milk emulsion. *J Food Eng [Internet].*, 2010 Apr [cited 2016 Jan 24], 97(3), 341–7. Available from: http://www.sciencedirect.com/science/article/pii/S0260877409005342.

[21] Onsaard, E; Vittayanont, M; Srigam, S; McClements, DJ. Properties and Stability of Oil-in-Water Emulsions Stabilized by Coconut Skim Milk Proteins. *J Agric Food Chem [Internet].*, 2005 Jul 1 [cited 2016 Jan 24], 53(14), 5747–53. Available from: http://dx.doi.org/10.1021/jf050312r.

[22] Dayrit, FM; Buenafe, OEM; Chainani, ET; de Vera, IMS; Dimzon, IKD; Gonzales, EG; et al., Essential quality parameters of commercial virgin coconut oil. *INDIAN COCONUT J-COCHIN- [Internet].*, 2007 [cited 2016 Jan 24], 38(5), 9. Available from: http://www.apccsec.org/CORD_ABSRTRACTS/Vol_23_1_2007/essential%20quality %20parameter......pdf.

[23] Puertollano, CL; Banzon, J; Steinkraus, KH. Separation of the oil and protein fractions in coconut (Cocos nucifera linn.) by fermentation. *J Agric Food Chem [Internet].*, 1970 Jul 1 [cited 2016 Jan 24], 18(4), 579–84. Available from: http://dx.doi.org/10.1021/jf60170a018.

[24] Hamid, MA; Sarmidi, MR; Mokhtar, TH; Sulaiman, WRW; Aziz, RA. Innovative integrated wet process for virgin coconut oil production. *J Appl Sci [Internet].*, 2011 [cited 2016 Jan 24], 11(13), 2467–9. Available from: http://www.researchgate.net/profile/Ramlan_Aziz/publication/252218780_Innovative_Integrated_Wet_Process_for_Virgin_Coconut_Oil_Production/links/0a85e53bc87397ded7000000.pdf.

[25] Chen, BK; Diosady, LL. Enzymatic aqueous processing of coconuts. *Int J Appl Sci Eng [Internet].*, 2003 [cited 2016 Jan 24], 1(1), 55–61. Available from: http://www.cyut.edu.tw/~ijase/2003/ijase_1(1)_5_55-61.pdf.

[26] Seow, CC; Gwee, CN. Coconut milk: chemistry and technology. *Int J Food Sci Technol [Internet].*, 1997 May 1 [cited 2016 Jan 24], 32(3), 189–201. Available from: http://onlinelibrary.wiley.com/doi/10.1046/ j.1365-2621.1997.00400.x/abstract.

[27] Rosenthal, A; Pyle, DL; Niranjan, K. Aqueous and enzymatic processes for edible oil extraction. *Enzyme Microb Technol [Internet].* 1996 Nov 1 [cited 2016 Jan 24], 19(6), 402–20. Available from: http://www.sciencedirect.com/science/article/pii/S014102299680004F.

[28] Nour, AH; Mohammed, FS; Yunus, RM; Arman, A. others. Demulsification of virgin coconut oil by centrifugation method: a feasibility study. *Int J Chem Technol [Internet].*, 2009 [cited 2016 Jan 29], 1(2), 59–64. Available from: http://docsdrive.com/pdfs/knowledgia/ijct/2009/59-64.pdf.

[29] Harni, M; Putri, SK. Processing Methode Effect to Virgin Coconut Oil (VCO) Quality After Storaging. *Int J Adv Sci Eng Inf Technol [Internet].*, 2014 [cited 2016 Jan 29], 4(2), 28–30. Available from: http://www.insightsociety.org/ojaseit/index.php/ijaseit/article/view/369.

[30] Songkro, S; Sirikatitham, A; Sungkarak, S; Buaking, K; Wungsintaweekul, J; Maneenuan, D; et al., Characterization of aromatherapy massage oils prepared from virgin coconut oil and some essential oils. *J Am Oil Chem Soc [Internet].*, 2010 [cited 2016 Jan 29], 87(1), 93–107. Available from: http://link.springer.com/article/10.1007/s11746-009-1465-5.

[31] Kamariah, L; Azmi, A; Rosmawati, A; Ching, MW; Azlina, MD; Sivapragasam, A; et al., Physico-chemical and quality characteristics of virgin coconut oil–A Malaysian survey. *J Trop Agric Fd Sc [Internet].*, 2008 [cited 2016 Jan 29], 36(2), 000–000. Available from: http://ejtafs.mardi.gov.my/jtafs/36-2/Virgin%20coconut%20oil.pdf.

[32] Fife, B. The Coconut Oil Miracle. *Penguin*, 2004. 260 p.

[33] St-Onge, MP; Jones, PJH. Physiological Effects of Medium-Chain Triglycerides: Potential Agents in the Prevention of Obesity. *J Nutr [Internet].*, 2002 Mar 1 [cited 2016 Jan 24], 132(3), 329–32. Available from: http://jn.nutrition.org/content/132/3/329.

[34] Assunção, ML; Ferreira, HS; dos Santos, AF; Cabral, CR; Florêncio, TMMT. Effects of dietary coconut oil on the biochemical and anthropometric profiles of women presenting abdominal obesity. *Lipids.*, 2009 Jul, 44(7), 593–601.

[35] Lim-Sylianco, CY. Anticarcinogenic effect of coconut oil. *Philipp J Coconut Stud.*, 1987, 12, 89–102.

[36] Yeap, SK; Beh, BK; Ali, NM; Yusof, HM; Ho, WY; Koh, SP; et al., Antistress and antioxidant effects of virgin coconut oil *in vivo*. *Exp Ther Med [Internet].*, 2015 [cited 2016 Jan 24], 9(1), 39–42. Available from: http://www.spandidos-publications.com/10.3892/etm.2014.2045? text= abstract.

[37] Lans, C. Comparison of plants used for skin and stomach problems in Trinidad and Tobago with Asian ethnomedicineJournal of Ethnobiology and Ethnomedicine. *J Ethnobiol Ethnomedicine [Internet].*, 2007 [cited 2016 Jan 29], 3, 3. Available from: http://www.biomedcentral.com/content/pdf/1746-4269-3-3.pdfNgameni.

[38] Agero, AL; Verallo-Rowell, VM. A randomized double-blind controlled trial comparing extra virgin coconut oil with mineral oil as a moisturizer for mild to moderate xerosis. *Dermat Contact Atopic Occup Drug [Internet].*, 2004 [cited 2016 Jan 29], 15(3), 109–16. Available from: http://europepmc.org/abstract/med/15724344.

[39] Zakaria, ZA; Rofiee, MS; Somchit, MN; Zuraini, A; Sulaiman, MR; The, LK; et al., Hepatoprotective activity of dried-and fermented-processed virgin coconut oil. *Evid Based Complement Alternat Med [Internet].*, 2011 [cited 2016 Jan 24], 2011. Available from: http://www.hindawi.com/journals/ecam/2011/142739/abs/.

[40] Dosumu, OO; Akinola, OB; Akang, EN. Alcohol-induced testicular oxidative stress and cholesterol homeostasis in rats–The therapeutic potential of virgin coconut oil. *Middle East Fertil Soc J [Internet].*, 2012 [cited 2016 Jan 29], 17(2), 122–8. Available from: http://www. sciencedirect.com/science/article/pii/S111056901100135X.

[41] Siddalingaswamy, M; Rayaorth, A; Khanum, F. Anti-diabetic effects of cold and hot extracted virgin coconut oil. *J Diabetes Mellit [Internet].*, 2011 [cited 2016 Jan 29], 1(04), 118. Available from: http://www.scirp.org/journal/PaperInformation.aspx?paperID=8421.

[42] Ogbolu, DO; Oni, AA; Daini, OA; Oloko, AP. *In vitro* antimicrobial properties of coconut oil on Candida species in Ibadan, Nigeria. *J Med Food.*, 2007 Jun, 10(2), 384–7.

[43] Yuniwarti, EYW; Asmara, W; Artama, WT; Tabbu, CR. The effect of Virgin Coconut Oil on lymphocyte and CD4 in chicken vaccinated against Avian Influenza virus. *J Indones Trop Anim Agric [Internet].*, 2012 [cited 2016 Jan 29], 37(1), 64–9. Available from: http://ejournal. undip.ac.id/index.php/jitaa/article/view/7476.

[44] What is Virgin Coconut Oil? [Internet]. [cited 2016 Jan 13]. Available from: http://www.tropicaltraditions.com/what_is_virgin_ coconut_oil. htm.

[45] how to choose vco and some production processes - production_processes_how_to_choose_vco.pdf [Internet]. [cited 2016 Jan 13]. Available from: http://www.cocoscience.com/pdf/production_ processes_how_to_choose_vco.pdf.

[46] Marina, AM; Man, YBC; Nazimah, S a. H; Amin, I. Antioxidant capacity and phenolic acids of virgin coconut oil. *Int J Food Sci Nutr.*, 2009, 60 Suppl 2, 114–23.

[47] Coconut oil: It's good for you after all [Internet]. [cited 2016 Jan 13]. Available from: http://www.sundaytimes.lk/111016/Plus/plus_05.html.

[48] Broschat, T; Crane, J. *The Coconut Palm in Florida [Internet].*, 2015 [cited 2016 Feb 3]. Available from: http://edis.ifas.ufl.edu/mg043.

[49] Li, X; Zhai, G; Gao, S; Shen, X. Decadal trends of global precipitation in the recent 30 years. *Atmospheric Sci Lett [Internet].*, 2015 Jan 1 [cited 2016 Feb 3], 16(1), 22–6. Available from: http://onlinelibrary.wiley. com/doi/10.1002/asl2.514/abstract.

[50] Singh, RH; Seepersad, G; Rankine, LB. The regional coconut industry: Global market intelligence [Internet]. *CARICOM Regional Transformation Programme for Agriculture*, 2007 Dec. Available from: http://www.caricom.org/jsp/community/agribusiness_forum/coconut_industry_market_intelligence.pdf.

[51] CRB Fundamentals [Internet]. [cited 2016 Jan 15]. Available from: http://www.crbtrader.com/fund/articles/coconut.asp.

[52] USDA. World Palm Oil, Coconut Oil, and Fish Meal Supply and Distribution [Internet]. 2016 Jan p. 24. Available from: http://apps.fas.usda.gov/psdonline/circulars/oilseeds.pdf.

[53] Microsoft Word - journal143_article03.doc - 17_1687.pdf [Internet]. [cited 2016 Jan 13]. Available from: http://fic.nfi.or.th/food/upload/pdf/17_1687.pdf.

[54] Microsoft Word - journal143_article03.doc - journal143_article03.pdf [Internet]. [cited 2016 Jan 13]. Available from: http://www.journal.au.edu/au_techno/2011/jan2011/journal143_article03.pdf.

[55] Coconut oil making "positive impact" on Solomon Islands economy - ABC News (Australian Broadcasting Corporation) [Internet]. [cited 2016 Jan 13]. Available from: http://www.abc.net.au/news/2015-07-18/coconut-oil-has-positive-impact-on-solomon-islands-economy/6630008.

[56] Fair Trade, USA. Launches Fair Trade Certified™ Coconuts | Fair Trade USA [Internet]. [cited 2016 Jan 13]. Available from: http://fairtradeusa.org/press-room/press_release/fair-trade-usa-launches-fair-trade-certified-coconuts.

[57] Robbins, SL; Kumar, V; Cotran, RS. Robbins and Cotran pathologic basis of disease. *Philadelphia, PA: Philadelphia, PA*, 2010.

[58] Seneviratne, KN; Sudarshana Dissanayake, DM. Variation of phenolic content in coconut oil extracted by two conventional methods. *Int J Food Sci Technol [Internet].*, 2008 Apr 1 [cited 2016 Jan 15], 43(4), 597–602. Available from: http://onlinelibrary.wiley.com/doi/10.1111/j.1365-2621.2006.01493.x/abstract.

[59] Intahphuak, S; Khonsung, P; Panthong, A. Anti-inflammatory, analgesic, and antipyretic activities of virgin coconut oil. *Pharm Biol [Internet].* 2010 Feb 1 [cited 2015 Oct 11], 48(2), 151–7. Available from:http://www.tandfonline.com/doi/abs/10.3109/13880200903062614.

[60] Vysakh, A; Ratheesh, M; Rajmohanan, TP; Pramod, C; Premlal, S; Girish kumar, B; et al., Polyphenolics isolated from virgin coconut oil inhibits adjuvant induced arthritis in rats through antioxidant and anti-inflammatory action. *Int Immunopharmacol.*, 2014 May, 20(1), 124–30.

[61] DiMartino, MJ; Jr, GKC; Wolff, CE; Hanna, N. The pharmacology of arachidonic acid-induced rat paw edema. *Agents Actions [Internet].*, 1987 Aug [cited 2016 Jan 15], 21(3-4), 303–5. Available from: http://link.springer.com/article/10.1007/BF01966498.

[62] Wan, JM; Grimble, RF. Effect of dietary linoleate content on the metabolic response of rats to Escherichia coli endotoxin. *Clin Sci Lond Engl*, 1979. 1987 Mar, 72(3), 383–5.

[63] Sadeghi, S; Wallace, FA; Calder, PC. Dietary lipids modify the cytokine response to bacterial lipopolysaccharide in mice. *Immunology.*, 1999 Mar, 96(3), 404–10.

[64] Dréno, B. Recent data on epidemiology of acne. *Ann Dermatol Vénéréologie [Internet].*, 2010 Dec [cited 2016 Jan 15], 137(12, Supplement 2), 3–5. Available from: http://www.sciencedirect.com/science/article/pii/S0151963810700454.

[65] Tan, JKL. Psychosocial impact of acne vulgaris: evaluating the evidence. *Skin Ther Lett.*, 2004 Sep, 9(7), 1–3, 9.

[66] Kim, J. Review of the Innate Immune Response in Acne vulgaris: Activation of Toll-Like Receptor 2 in Acne Triggers Inflammatory Cytokine Responses. *Dermatology*

[Internet]., 2005 [cited 2016 Jan 15], 211(3), 193–8. Available from: http://www.karger.com/doi/10. 1159/000087011.

[67] Nakatsuji, T; Kao, MC; Fang, JY; Zouboulis, CC; Zhang, L; Gallo, RL; et al., Antimicrobial Property of Lauric Acid Against Propionibacterium Acnes: Its Therapeutic Potential for Inflammatory Acne Vulgaris. *J Invest Dermatol [Internet]*., 2009 Apr 23 [cited 2016 Jan 27], 129(10), 2480–8. Available from: http://www.nature.com/jid/journal/v129/n10/abs/jid200993a.html.

[68] Chun, J; Lee, J; Ye, L; Exler, J; Eitenmiller, RR. Tocopherol and tocotrienol contents of raw and processed fruits and vegetables in the United States diet. *J Food Compos Anal [Internet]*., 2006 [cited 2016 Jan 24], 19(2), 196–204. Available from: http://www.sciencedirect.com/science/article/pii/S0889157505000967.

[69] Jordan, M; Nayel, A; Brownlow, B; Elbayoumi, T. Development and evaluation of tocopherol-rich argan oil-based nanoemulsions as vehicles possessing anticancer activity. *J Biomed Nanotechnol [Internet]*. 2012 [cited 2016 Jan 24], 8(6), 944–56. Available from: http://www.ingentaconnect.com/content/asp/jbn/2012/00000008/00000006/art0000.

[70] Winarsi, H; HERNAYANTI, H; PURWANTO, A. Virgin coconut oil (VCO) enriched with Zn as immunostimulator for vaginal Candidiasis patient. *HAYATI J Biosci [Internet]*., 2009 [cited 2016 Jan 24], 15(4), 135. Available from: http://jesl.journal.ipb.ac.id/index.php/hayati/article/viewArticle/133.

[71] Kappally, S; Shirwaikar, A; Shirwaikar, A. COCONUT OIL–A REVIEW OF POTENTIAL APPLICATIONS. 2015 [cited 2016 Jan 24]; Available from: http://www.hygeiajournal.com/Downloads/1075302624r-3%20Coconut%20Oil%20revised(2).pdf.

[72] Toth, PP. The "Good Cholesterol" High-Density Lipoprotein. Circulation [Internet]. 2005 Feb 8 [cited 2016 Feb 1], 111(5), e89–91. Available from: http://circ.ahajournals.org/content/111/5/e89.

[73] Nevin, KG; Rajamohan, T. Beneficial effects of virgin coconut oil on lipid parameters and *in vitro* LDL oxidation. *Clin Biochem*., 2004 Sep, 37(9), 830–5.

[74] Nevin, KG; Rajamohan, T. Virgin coconut oil supplemented diet increases the antioxidant status in rats. *Food Chem [Internet]*., 2006 [cited 2016 Jan 14], 99(2), 260–6. Available from: http://www.sciencedirect.com/science/article/pii/S0308814605006412.

[75] HARINI, M; PARAMA ASTIRIN, O. Blood cholesterol levels of hypercholesterolemic rat (Rattus norvegicus) after VCO treatment. *Nusant Biosci [Internet]*., 2009 [cited 2016 Jan 24], 1(2). Available from: http://jurnal.pasca.uns.ac.id/index.php/ nubios/article/view/28.

[76] Nevin, KG; Rajamohan, T. Influence of virgin coconut oil on blood coagulation factors, lipid levels and LDL oxidation in cholesterol fed Sprague–Dawley rats. Eur E-J Clin Nutr Metab [Internet]. 2008 Feb 1 [cited 2015 Oct 11], 3(1), e1–8. Available from: http://e-spenjournal.org.marlin-prod.literatumonline.com/article/S1751499107000431/abstract.

[77] Nevin, KG; Rajamohan, T. Wet and dry extraction of coconut oil: impact on lipid metabolic and antioxidant status in cholesterol coadministered rats. *Can J Physiol Pharmacol [Internet]*., 2009 [cited 2016 Jan 24], 87(8), 610–6. Available from: http://www. nrcresearchpress.com/doi/abs/10.1139/y09-045.

[78] Seneviratne, KN; HapuarachchI, CD; Ekanayake, S. Comparison of the phenolic-dependent antioxidant properties of coconut oil extracted under cold and hot conditions. *Food Chem [Internet].*, 2009 [cited 2016 Jan 24], 114(4), 1444–9. Available from: http://www.sciencedirect.com/science/article/pii/S0308814608013745.

[79] Norulaini, NN; Setianto, WB; Zaidul, ISM; Nawi, AH; Azizi, CYM; Omar, AM. Effects of supercritical carbon dioxide extraction parameters on virgin coconut oil yield and medium-chain triglyceride content. *Food Chem [Internet].*, 2009 [cited 2016 Jan 24], 116(1), 193–7. Available from: http://www.sciencedirect.com/science/article/pii/S030881 4609002179.

[80] Villarino, BJ; Dy, LM; Lizada, MCC. Descriptive sensory evaluation of virgin coconut oil and refined, bleached and deodorized coconut oil. LWT-Food Sci Technol [Internet]. 2007 [cited 2016 Jan 24], 40(2), 193–9. Available from: http://www.sciencedirect.com/science/article/pii/S0023643805002495.

[81] Abujazia, MA; Muhammad, N; Shuid, AN; Soelaiman, IN. The effects of virgin coconut oil on bone oxidative status in ovariectomised rat. *Evid-Based Complement Altern Med ECAM.*, 2012, 2012, 525079.

[82] Hayatullina, Z; Muhammad, N; Mohamed, N; Soelaiman, IN. Virgin coconut oil supplementation prevents bone loss in osteoporosis rat model. *Evid-Based Complement Altern Med ECAM.*, 2012, 2012, 237236.

[83] Arunima, S; Rajamohan, T. Effect of virgin coconut oil enriched diet on the antioxidant status and paraoxonase 1 activity in ameliorating the oxidative stress in rats–a comparative study. Food Funct [Internet]. 2013 [cited 2016 Jan 24], 4(9), 1402–9. Available from: http://pubs.rsc.org/en/content/articlehtml/2013/fo/c3fo60085h.

[84] Otuechere, CA; Madarikan, G; Simisola, T; Bankole, O; Osho, A. Virgin coconut oil protects against liver damage in albino rats challenged with the anti-folate combination, trimethoprim-sulfamethoxazole. J Basic Clin Physiol Pharmacol [Internet]. 2014 [cited 2016 Jan 24], 25(2), 249–53. Available from: http://www.degruyter.com/view/j/jbcpp.2014. 25. issue-2/jbcpp-2013-0059/jbcpp-2013-0059.xml.

[85] Zakaria, ZA; Somchit, MN; Mat Jais, AM; The, LK; Salleh, MZ; Long, K. *In vivo* antinociceptive and anti-inflammatory activities of dried and fermented processed virgin coconut oil. *Med Princ Pract [Internet].*, 2011 [cited 2016 Jan 24], 20(3), 231–6. Available from: http://www.karger.com/Article/Fulltext/323756.

[86] Eid, HH; Labib, RM; Hamid, NSA; Hamed, MA; Ross, SA. Hepatoprotective and antioxidant polyphenols from a standardized methanolic extract of the leaves of Liquidambar styraciflua L. *Bull Fac Pharm Cairo Univ [Internet].*, 2015 Dec [cited 2016 Jan 24], 53(2), 117–27. Available from: http://www.sciencedirect.com/science/article/pii/S1110093115000265.

[87] Enig, MG. Coconut: in support of good health in the 21st century. *36th Sess Asian Pac Coconut Community APCC Singap [Internet].*, 1999 [cited 2016 Jan 14]; Available from: http://www.snc.sg/about-us/publications/coconut-oil/coconut-in-support-of-good-health-in-the-21st-century.

[88] Nurul-Iman, BS; Kamisah, Y; Jaarin, K; Qodriyah, HMS. Virgin coconut oil prevents blood pressure elevation and improves endothelial functions in rats fed with repeatedly heated palm oil. *Evid-Based Complement Altern Med ECAM.*, 2013, 2013, 629329.

[89] Kamisah, Y; Periyah, V; Lee, KT; Noor-Izwan, N; Nurul-Hamizah, A; Nurul-Iman, BS; et al., Cardioprotective effect of virgin coconut oil in heated palm oil diet-induced hypertensive rats. *Pharm Biol.*, 2015, 53(9), 1243–9.

[90] Subermaniam, K; Saad, QHM; Bakhtiar, S; Hamid J; Sidek, F; Othman, F. Effects of Virgin Coconut Oil on the Histomorphometric Parameters in the Aortae and Hearts of Rats Fed with Repeatedly Heated Palm Oil. *International Journal of Bioscience, Biochemistry and Bioinformatics.*, 2015, 5(2), 120–31.

[91] Liau, KM; Lee, YY; Chen, CK; Rasool, AHG. An open-label pilot study to assess the efficacy and safety of virgin coconut oil in reducing visceral adiposity. *ISRN Pharmacol.*, 2011, 2011, 949686.

[92] Feranil, AB; Duazo, PL; Kuzawa, CW; Adair, LS. Coconut oil is associated with a beneficial lipid profile in pre-menopausal women in the Philippines. *Asia Pac J Clin Nutr.*, 2011, 20(2), 190–5.

[93] ChiawMei, S; HipSeng, Y; ChoonMei, L. Commercial virgin coconut oil: assessment of antimicrobial potential. *Asian J Food Agro-Ind.*, 2010, 3(6), 567–79.

[94] Oyi, A; Onaolapo, J; Obi, R. Formulation and Antimicrobial Studies of Coconut (Cocos nucifera Linne) Oil - Open Access Library. *Research Journal of Applied Sciences, Engineering, and Technology [Internet].*, 2010 [cited 2016 Jan 14], 2(2), 133–7. Available from: http://www. jourlib.org/paper/2569079#.VpcflFLFwak.

[95] Shankar, P; Ahuja, S; Tracchio, A. Agro FOOD Industry Hi Tech articles [Internet]. Coconut oil: A review. 2013 [cited 2016 Jan 14]. Available from: http://www. teknoscienze.com/articles/agro-food-industry-hi-tech-coconut-oil-a-review.aspx#.Vpcg DlLFwak.

[96] Thormar, H; Hilmarsson, H. The role of microbicidal lipids in host defense against pathogens and their potential as therapeutic agents. *Chem Phys Lipids.*, 2007 Nov, 150(1), 1–11.

[97] Verallo-Rowell, VM; Dillague, KM; Syah-Tjundawan, BS. Novel antibacterial and emollient effects of coconut and virgin olive oils in adult atopic dermatitis. *Dermat Contact Atopic Occup Drug.*, 2008 Dec, 19(6), 308–15.

[98] Evangelista, MTP; Abad-Casintahan, F; Lopez-Villafuerte, L. The effect of topical virgin coconut oil on SCORAD index, transepidermal water loss, and skin capacitance in mild to moderate pediatric atopic dermatitis: a randomized, double-blind, clinical trial. *Int J Dermatol [Internet].*, 2014 Jan 1 [cited 2016 Jan 14], 53(1), 100–8. Available from: http://onlinelibrary.wiley.com/doi/10.1111/ijd.12339/abstract.

[99] Tangwatcharin, P; Khopaibool, P. Activity of virgin coconut oil, lauric acid or monolaurin in combination with lactic acid against Staphylococcus aureus. *Southeast Asian J Trop Med Public Health.*, 2012 Jul, 43(4), 969–85.

[100] Shilling, M; Matt, L; Rubin, E; Visitacion, MP; Haller, NA; Grey, SF; et al., Antimicrobial effects of virgin coconut oil and its medium-chain fatty acids on Clostridium difficile. *J Med Food.*, 2013 Dec, 16(12), 1079–85.

[101] Silalahi, J; Permata, Y; Putra, ED. lux. ANTIBACTERIAL ACTIVITY OF HYDROLYZED VIRGIN COCONUT OIL. *Asian J Pharm Clin Res [Internet].*, 2014 May 17 [cited 2016 Jan 14], 7(3). Available from: http://innovareacademics.in/journals/ index.php/ajpcr/article/view/1042.

[102] Elysa, Harahap, U; Silalahi, J. Antibacterial activity of enzymatic hydrolysis of Virgin Coconut Oil against Salmonella. *Int J PharmTech Res.*, 2014, 6(2), 589–99.

[103] DebMandal, M; Mandal, S. Coconut (Cocos nucifera L.: Arecaceae): In health promotion and disease prevention. *Asian Pac J Trop Med [Internet].*, 2011 Mar [cited 2016 Jan 14], 4(3), 241–7. Available from: http://www.sciencedirect.com/science/article/pii/S1995764511600783.

[104] Arora, R; Chawla, R; Marwah, R; Arora, P; Sharma, RK; Kaushik, V; et al., Potential of Complementary and Alternative Medicine in Preventive Management of Novel H1N1 Flu (Swine Flu) Pandemic: Thwarting Potential Disasters in the Bud. *Evid-Based Complement Altern Med ECAM.*, 2011, 2011, 586506.

[105] Geddes, L. Are diabetes, dementia and depression triggered by one SILENT KILLER? [Internet]. Mail Online. 2015 [cited 2016 Feb 4]. Available from: http://www.dailymail.co.uk/health/article-3125357/Furred-arteries-Diabetes-Eye-disease-Dementia-Depression-doctors-believe-triggered-one-SILENT-KILLER.html.

[106] Olmos-Alonso, A; Schetters, STT; Sri, S; Askew, K; Mancuso, R; Vargas-Caballero, M; et al., Pharmacological targeting of CSF1R inhibits microglial proliferation and prevents the progression of Alzheimer's-like pathology. *Brain [Internet].* 2016 Jan 8 [cited 2016 Feb 4], awv379. Available from: http://brain.oxfordjournals.org/content/early/2016/01/07/brain.awv379.

[107] Hu Yang, I; De la Rubia Ortí, JE; Selvi Sabater, P; Sancho Castillo, S; Rochina, MJ; Manresa Ramón, N; et al., [COCONUT OIL: NON-ALTERNATIVE DRUG TREATMENT AGAINST ALZHEIMER´S DISEASE]. *Nutr Hosp.*, 2015, 32(6), 2822–7.

[108] Gandotra, S; Kour, J; Van der Waag, A. Efficacy of Adjunctive Extra Virgin Coconut Oil Use in Moderate to Severe Alzheimer? s Disease. *Int J Sch Cogn Psychol [Internet].*, 2014 [cited 2016 Feb 4], 2014. Available from: http://www.omicsonline.com/open-access/efficacy-of-adjunctive-extra-virgin-coconut-oil-use-in-moderate-to-severe-alzheimers-disease--1234-3425.1000108.php?aid=27142.

[109] Narayanankutty, A; Mukesh, RK; Ayoob, SK; Ramavarma, SK; Suseela, IM; Manalil, JJ; et al., Virgin coconut oil maintains redox status and improves glycemic conditions in high fructose fed rats. *J Food Sci Technol.*, 2016 Jan, 53(1), 895–901.

[110] Brenna, JT; Kothapalli, KSD. Commentary on "Influence of virgin coconut oil-enriched diet on the transcriptional regulation of fatty acid synthesis and oxidation in rats--a comparative study" by Sakunthala Arunima and Thankappan Rajamohan. *Br J Nutr.*, 2014 Nov 14, 112(9), 1425–6.

INDEX

B

C

D

G

H

J

K

L

M

N

O

P

Q

R

T